Food Product Development

Food Product Development

Avantina Sharma PhD
Nutritionist

CBS

CBS Publishers and Distributors Pvt Ltd

New Delhi • Bengaluru • Chennai • Kochi • Kolkata • Mumbai
Hyderabad • Jharkhand • Nagpur • Patna • Pune • Uttarakhand

Disclaimer

Science and technology are constantly changing fields. New research and experience broaden the scope of information and knowledge. The author has tried her best in giving information available to her while preparing the material for this book. Although all efforts have been made to ensure optimum accuracy of the material, yet it is quite possible some errors might have been left uncorrected. The publisher, the printer and the author will not be held responsible for any inadvertent errors or inaccuracies.

Food Product Development

ISBN: 978-93-86827-95-1

First Edition: 2018

Reprint 2021

Published by Satish Kumar Jain and produced by Varun Jain for

CBS Publishers & Distributors Pvt Ltd

4819/XI Prahlad Street, 24 Ansari Road, Daryaganj, New Delhi 110 002, India.
Ph: 23289259, 23266861, 23266867   Website: www.cbspd.com
Fax: 011-23243014   e-mail: delhi@cbspd.com; cbspubs@airtelmail.in.

Corporate Office: 204 FIE, Industrial Area, Patparganj, Delhi 110 092
Ph: 4934 4934   Fax: 4934 4935   e-mail: publishing@cbspd.com; publicity@cbspd.com

Branches

• **Bengaluru:** Seema House 2975, 17th Cross, K.R. Road, Banasankari 2nd Stage, Bengaluru 560 070, Karnataka, India
Ph: +91-80-26771678/79   Fax: +91-80-26771680   e-mail: bangalore@cbspd.com

• **Chennai:** 7, Subbaraya Street, Shenoy Nagar, Chennai 600 030, Tamil Nadu, India
Ph: +91-44-26680620, 26681266   Fax: +91-44-42032115   e-mail: chennai@cbspd.com

• **Kochi:** Ashana House, No. 39/1904, AM Thomas Road, Valanjambalam, Ernakulam 682 016, Kochi, Kerala, India
Ph: +91-484-4059061-65   Fax: +91-484-4059065   e-mail: kochi@cbspd.com

• **Kolkata:** 6/B, Ground Floor, Rameswar Shaw Road, Kolkata-700 014, West Bengal, India
Ph: +91-33-22891126, 22891127, 22891128   e-mail: kolkata@cbspd.com

• **Mumbai:** PWD Shed, Gala No. 25/26, Ramchandra Bhatt Marg, Next JJ Hospital, Gate No. 2, Opp. Union Bank of India, Noorbaug, Mumbai-400009, Maharashtra, India
Ph: +91-22-66661880/89   e-mail: mumbai@cbspd.com

Representatives

• **Hyderabad**	0-9885175004	• **Jharkhand**	0-9811541605	• **Nagpur**	0-9421945513
• **Patna**	0-9334159340	• **Pune**	0-9623451994	• **Uttarakhand**	0-9716462459

Printed at India Binding House, Noida, UP, India

to

**RK Bhandari, Meena Bhandari, Mahendra Bhandari
and Sushma Bhandari**

for their inspiration

Mahendra Singhwi and Manju Singhwi

for teaching me the passion for creativity and innovation

Saurabh Bhandari

for being my anchor

Preface

There has been a gigantic leap in the food products sector in the last few decades. Given the fact that these days many youngsters are taking the path of entrepreneurship coupled with a universal love of food, it becomes important to study product development in the context of food.

Product development is a multi-disciplinary activity and this book introduces the readers to a systematic process which integrates the various research areas, and which identifies the activities, outcomes and decisions to be made as the food product development project progresses.

The book also aims to incorporate some of the ideas from other industries such as product concept engineering and product design, so that these can be considered by management in the food industry. The book has been developed as an interactive text so that it can be used both in the food industry as a teaching method for people starting to work in product development and also as a textbook for an introductory course in product development for undergraduate as well as postgraduate students.

The book helps the readers from their journey of idea conceptualisation to seeing it physically before their eyes. The book discusses all nuances related to launching a new or modified food product into the market.

I hope the readers will find a treasure trove of information in the book which would benefit them in their future endeavours.

Best wishes to the student community, entrepreneurs, food enthusiasts and interested readers.

Avantina Sharma

Contents

Preface   *vii*

1. Food Product Development: Concept to Concrete   1

2. Market and Consumer Research   18

3. Determinants of Food Choices:
   Trends in Social Change   41

4. Food Consumption Trends   52

5. Traditional Foods: Shifting Trends, Patterns and
   Innovations   73

6. Food Processing   86

7. Convenience Foods   102

8. Fast Food   122

9. Quick Cooking Products   134

10. Food Additives and Preservatives   150

11. Standardization of Recipes   180

12. Food Packaging, Graphics and Labelling   194

13. Transportation of Food Products   241

14. Sensory Evaluation and Quality Control   260

15. Extrepreneurship   302

16. Equipment   326

17. Setting a Price for the Product   341

18. Advertising and Marketing   349

19. Food Laws   366

20. Post-harvest Losses and Technology   375

Bibliography   391

Index   397

Chapter 1

Food Product Development: Concept to Concrete

New product development (NPD) is the term used to describe the complete process of bringing a new product or service to market.

There are two parallel paths involved in the NPD process:
1. Idea generation, product design, and detail engineering
2. Market research and marketing analysis.

During product development, one finds various levels of products like: (the concept depicted in Fig. 1.1)
- New to the market
- New to the company
- Completely novel and create totally new markets
- New product concepts which are merely minor modifications of existing products
- New products which are completely innovative to the company.

Stages in Product Development

1. **Idea generation**
   a. Ideas for new products can be obtained from consumers, the R and D department of the enterprise, competitors, focus groups, employees, salespeople, corporate spies, trade shows, or through a policy of open innovation.
   b. Formal idea generating techniques include attribute listing, brainstorming, morphological analysis, problem analysis.

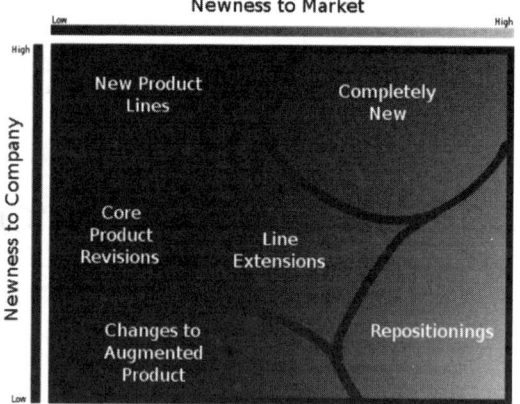

Fig. 1.1: Types of product innovations

2. **Idea screening**
   a. The object is to eliminate unsound concepts prior to devoting resources to them.
   b. The screeners must ask at least three questions:
      i. Will the customer in the target market benefit from the product?
      ii. Is it technically feasible to manufacture the product?
      iii. Will the product be profitable when manufactured and delivered to the customer at the target price?
3. **Concept development and testing**
   a. Develop the marketing and engineering details
      i. Who is the target market and who is the decision maker in the purchasing process?
      ii. What product features must the product incorporate?
      iii. What benefits will the product provide?
      iv. How will consumers react to the product?
      v. How will the product be produced most cost effectively?
      vi. Prove feasibility through virtual computer aided rendering and rapid prototyping.
      vii. What will it cost to produce it?

    b. Test the concept by asking a sample of prospective customers what they think of the idea.

4. **Business analysis**
   a. Estimate likely selling price based upon competition and customer feedback.
   b. Estimate sales volume based upon size of market.
   c. Estimate profitability and break even point.

5. **Beta testing and market testing**
   a. Produce a physical prototype or mock-up.
   b. Test the product in typical usage situations.
   c. Conduct focus group customer interviews or introduce at trade show.
   d. Make adjustments where necessary.
   e. Produce an initial run of the product and sell it in a test market area to determine customer acceptance.

6. **Technical implementation**
   a. New program initiation.
   b. Resource estimation.
   c. Requirement publication.
   d. Engineering operations planning.
   e. Department scheduling.
   f. Supplier collaboration.
   g. Resource plan publication.
   h. Program review and monitoring.
   i. Contingencies—what-if planning.

7. **Commercialization**
   a. Launch the product.
   b. Produce and place advertising and other promotions.
   c. Fill the distribution (business) pipeline with product.
   d. Critical path analysis is most useful at this stage.

Food product development is the process of making new or modified food products. The process of product development involves a complex series of stages, requiring the combined talents of many specialists to make it successful.

Product development is by definition a future-oriented practice. It is an effort to foresee the future needs of the market place and to translate this information into state-of-the-art products.

In the present market scenario of processed foods, one observes unpredictable fluctuations in sales profile of established brand names. The once popular products may lose its premium position in the market prompting product innovation and strategy shift. One such strategy is to introduce new products having better consumer appeal.

In order to develop a new product, the manufacturer should have the knowledge about all the factors influencing the development of new products.

1. Product developers should be familiar with marketing, financial implications and logistics of introducing a new product into the competitive market.
2. They have to know how to undertake product development work from the customer's perspective.
3. Products will have to be developed in response to a specific, clearly identified market opportunities.
4. In addition to consumer demands, the driving forces behind new product development are:
   a. Peer competition,
   b. Availability of new technology,
   c. Alternate raw materials,
   d. Desire for novelty.
5. Market fragmentation leads to developing different types of foods for various categories of consumers.
6. Increase in the purchasing power of sections of population
7. Changing lifestyles
8. Increased emphasis on health and nutrition put pressure on processors to offer something new to the consumer.
9. Science and technology in the food industry are increasingly defined by the demands and perception of the market place.
   o To cope with such a situation, the food industry has to resort to product development and innovation as a continuous activity.

Specifically talking about FPD, below are the stages for the process:

1. Develop ideas for a new product.
2. Test ideas on a small scale.

3. Preliminary sensory evaluation.
4. Modify product.
5. Pilot plant.
6. Advanced sensory evaluation.
7. Perform consumer testing.
8. Finalise product specifications.
9. Manufacture product on a large scale.
10. Advertise the product.
11. Launch new product.

When product development is specifically aimed at needs of the consumers and attracting their attention towards buying and using a new product the following are kept in mind.

Consumer oriented approach consists of five steps:

1. **Identify the opportunity for new product:** Opportunity identification stage concerns the definition of the best market segment to introduce the new product and the generation of new product ideas.

2. **Design of the new product:** At the product design phase, the plan of action is:
   a. To identify the key benefits the product is to provide to the consumer.
   b. Positioning of these benefits versus competitive products.
   c. Development of the product and marketing strategy.
   d. Purchasing pattern of the consumer, what they consider important in their choice behaviour and what they consider shortcomings in present product supply provide an insight into some basic issues of product development.
   e. Consumer perception and preference have to be carefully studied while formulating a new product.

3. **Market testing of the new product:** Market testing may involve segmentation to identify relevant subgroups of consumers that are homogeneous in terms of preference and purchasing behaviour. If the new product is promising, as judged by the testing, it is ready for commercial sale.

4. **Launching the product in identified markets**
5. **Life-cycle management.**

Stages in Food Product Formulation

The process of developing a new product comprises three phases.

1. Formulation or recipe development
2. Standardization of processing methods
3. Evaluation and testing of quality parameters of the final product.

The important issues pertaining to new food product development are to take all measures in order to make the product successful. Some of the points of vital importance are:

- ○ Researching an optimized recipe to work on:
  - Most of the food products contain a variety of ingredients and preparing the product involves several processing steps.
  - Therefore working on a new product one has to study whether the recipes and methods of processing can yield an acceptable product or is there a need for modifying them to improve the quality and to meet the demands of the market.
- ○ Working on effective processing conditions.
- ○ Procuring food products of high quality.
- ○ Making accurate prediction, through research, of marketability.

Methods and Methodology adopted for Product Development

1. **One-factor-at-a time method or trial and error method:** A simple method where only one of the ingredients, in a formulation, is changed at a time.

   For instance, in potato chips where flavourings, salt and sugar are the vital ingredients considered as the variables which influence the product quality. Alterations are made in one ingredient while keeping the other two constant to achieve an optimal level, whereby the others are adjusted the same way. The optimal levels of flavourings. Salt and sugar are then combined in the preparation which will have the overall optimum quality. The trial and error method has been in use for a long time.

But this optimization method has certain disadvantages. These include:

1. It is laborious and time consuming,
2. It does not provide information about variable interaction effects, and
3. Achieved optimum consists only one variable levels that are actually tested.

Statistical Experimental Methods

Given the shortcomings of the above method, statistical experimental methods are used in product development for greater accuracy. Some helpful terms are:

1. *Independent variables* also known as *factors* are the parameters or characteristics, including ingredients and processing conditions, which have an effect on product quality.
2. *Dependent variables,* or the *responses* are the important measurable food quality indices. These are influenced directly or indirectly by different factors. Some examples of responses are sensory quality, nutritional value, chemical composition, microbiological characteristics and shelf-life.
3. *Test levels* or *levels* are the quantity of factors selected to be tested in the experimental design. A combination of factor levels is chosen according to the experimental design.
4. A *model* is a mathematical equation that describes the relationship between the response values and different factors quantitatively. It can predict optimized combination of factors to obtain products having required quality.

While dealing with the experimental design method, different kinds of system problems are encountered.

Product Process Problem

In food product design, there are two different kinds of system problems:

Process Problems

In the process problem, all the independent variables are not related to each other but are *orthogonal* to each other.

The change of one variable is not restricted by another variable.

In bread making, e.g. the temperature of the baking oven can, in principle, be chosen without any influence on the setting of baking time.

Mixture Problems or Recipe Problem

Recipe is one of the most important factors leading to successful food products. A recipe usually includes several ingredients, which have different effects on specific food quality. To study these effects is the prerequisite for being able to choose the optimal recipes any food products are manufactured by mixing two or more ingredients.

In bread, e.g. flour, sugar, baking powder, shortening and water are used. In this case, one or more properties of the food product generally depend only on the proportions of the ingredients present in the mixture and not on the amount of the mixture. One ingredient (an independent variable) cannot vary without changing at least one of the other ingredients in the mixture, because all the ingredients will be a part of a constant sum of 100%.

Troubleshooting

As described above, the distinguishing feature of a mixture problem is that the independent or controllable factors represent proportionate amounts of the mixture rather than unrestrained amounts. These proportions are measured by volume, by weight, or by mole fraction. These are non-negative numbers, and if expressed as fractions of the mixture, they must add up to a unity, especially if the ingredients to be studied are the only ingredients comprising the mixture.

Generally, all problems that appear in food product design can be divided into mixture or process problems, with the latter having the dominant share. Sometimes a problem that seems to be a mixture problem is really a process problem and can only be solved with a corresponding factorial experimental method.

In practice, it is not easy to distinguish a process problem with a mixture problem, when the food product design is only concerned with recipe or formulation development.

In order to address these problems certain measures are used as described below.

A factorial experiment: It studies the effect of some independent variables on food quality indices (response) through varying two or more of these independent variables, such as temperature, time, pressure and pH value. A series of values or test levels of each factor is selected and certain combinations of their levels are tested.

A mixture experiment: An experiment in which the food quality indices (response) are assumed to depend only on the relative proportions of the ingredient components present in the mixture and not on the amount of the mixture. In such an experiment, if the total amount of the mixture is held constant, the value of the response changes when changes are made in the relative proportions of the ingredients.

The Step-wise Detailed Explanation of Food Product Development Strategy

Traditionally, there have been six key stages in the process of new product development:
1. Genesis
2. Preliminary evaluation
3. Early development
4. Advanced development
5. Introduction
6. In-market evaluation

Today, some companies are emphasizing shorter developmental timelines and "getting to market" more quickly. In these cases, the development process may be considerably changed, consisting of only four stages:
1. Genesis and evaluation
2. Early development
3. Introduction
4. In-market evaluation and advanced development.

The Idea

New products ideas generally emerge from these organizational functions:

- o Marketing management, addressing business needs
- o Research and development.

Marketing Management

Marketing management personnel are involved in a continuous examination of changing consumer needs, developing market opportunities, and competitive developments.

Consumer research frequently is employed in this process to track consumer behaviour, which summarizes purchasing patterns and trends, as well as consumer needs and issues.

Research and Development

Depending on the capacities of an individual company, its R and D function may be engaged in primary research, which involves a technological development or breakthrough that suggests a new product; or it may be engaged in exploring "what is possible" within an area that has already been identified as important, or within an area that is of marketing interest to the company.

THE DEVELOPMENT OF THE IDEA

From Idea to Concept

Once an idea is conceived, it is important to quickly translate it into a concept that is intelligible to consumers. An initial concept statement is generally a paragraph long and clearly states:

- o The product idea.
- o Key product benefits.
- o What is unique or different about the product.

The Evaluation Process

Using qualitative insights and judgment, the new product development team reaches a point at which it is satisfied that the concept effectively communicates what the new product is and what its key benefits are.

The shift from development to evaluation is an important one. Generally, quantitative research methods will begin to be employed now, although the use of qualitative methods and judgment often continues to be an important part of the process. It is useful to separate the evaluation process into two phases: Preliminary evaluation and volumetric estimation.

Preliminary quantitative evaluation utilizes various approaches, often called concept screening or concept testing. Exposure to prototype products may or may not be employed, depending on whether exposure to an actual product is essential to an understanding of concept appeal, and whether it is practically possible to include the product.

Typically, concept testing or screening will examine:

1. The absolute appeal of the new product concept. This is often called a monadic rating, in which the concept is not compared to other concepts or products; intrinsic concept appeal itself is measured.

2. The comparative appeal of the new product concept to other new product concepts and existing products. This comparative measure may be used to screen a number of potential ideas and select the most promising concept, or it may assess the appeal of individual new product concepts relative to proven concepts or existing products.

The speed with which the evaluation process takes place varies dramatically between products and companies. Today, there is considerable discussion of approaches that get to market quickly. Management judgement, R and D, advertising, and manufacturing may be actively involved in exploring technical developmental issues during the evaluation process or may not continue to be active after the go/no go decision is made. Generally, time and money will be closely monitored until a go decision is reached.

At this point, a number of activities involving consumer research generally occur simultaneously. They include advertising development, name development, packaging development, pricing, and product optimization.

Advertising Development

Advertising development can be divided into three phases:
1. Early development
2. Initial approaches
3. Advanced development.

The process of concept development is the preliminary or early development stage of advertising development. In this phase, advertising is at the idea point; the advertising agency or the communications consultant may explore a wide variety of preliminary advertising ideas.

During the early phase, the level of commitment to a specific direction generally is low. Advertising agency account and creative personnel are in a generative mode, considering many communications issues and opportunities.

Simultaneously, other team members are focusing on the viability of the new product idea.

By the time of the second stage initial approaches, the advertising agency has identified as few as two or as many as ten potential directions for the advertising to take. If television advertising is being considered, the agency frequently will have translated these approaches into story boards, which combine initial copy efforts with artistic drawings. Story boards also may be animated with sound and action via videotape, and exposed as animatics.

In the third stage of advanced development, the agency generally has settled on a specific approach. This approach positions the product both in the marketplace and in the consumer's mind and thus is referred to as its positioning. If television is to be used, the agency often has committed to actual filming. At this point, considerable expenditures have been made, and commitment to a given direction is high.

Once a commercial is produced, it may or may not be quantitatively assessed before being aired on television.

Packaging Development

The role of consumer research in packaging development varies markedly. In some cases, consumers are not directly involved; in others, the range of qualitative and quantitative methods discussed above may be employed.

In concert with meeting functional criteria (i.e. the package contains and protects the product; the package is portable, storable, and easy to open) the package must communicate the product's brand and usage and must differentiate it from the competition. Consumer research also may consider the logo and its ability to convey the concept or positioning.

When there is sufficient time, the packaging may be evaluated qualitatively in the focus group interview, where consumers may consider the category of products, the competitive environment, and the package design's ability to generate impact and trial interest within that setting.

In another research format, a tachistoscope (T-scope) may be employed, which systematically exposes individual consumers to projected images of packaging designs and ascertains their ability to notice, discriminate, and retain packaging information.

A package's impact is read by determining which alternative design communicates the appropriate information and imagery in the least amount of time.

Name Development

The development of product names is a gray area of consumer research. No hard or fast methods for name development or evaluation have been devised. Indeed, the process of name development frequently is contentious and problematic.

Consumer research, both qualitative and quantitative, can be used to set parameters for name development, outlining criteria and characteristics that may be relevant. Researchers can, for instance, suggest that the name may need to emphasize a key product benefit (e.g. low calorie products with the name Lean Cuisine). The name must also be easy to pronounce and to remember, have no negative associations, and fit the positioning of the product.

Name consultants who use linguistic techniques can take a structured approach to discover how consumers use and understand language, as well as which sounds will best communicate name appeal.

Consumer research can be useful in name evaluation, although it is a complicated and involved process. Generally, research can reveal whether a name fulfills the criteria that were defined as essential for its success prior to its development, not whether the name is empirically "good" or "bad."

A particularly frustrating aspect of name development is the extent to which names have been previously trademarked.

Pricing

Pricing decisions frequently are dictated by product costs and objectives and competitive pricing levels.

From a consumer perspective, price sensitivity is highly variable.

As a rule, the more innovative the product, the less capable consumers are of evaluating its price and value.

Consumers weigh the value of each product and make appropriate trade-offs. They may pay more for a product that they perceive to be of high quality or prestigious.

As a rule, qualitative research is not a reliable method for examining pricing issues. However, a qualitative framework for pricing can be determined by examining consumers' motivations to buy, attitudes, geographical considerations, and category perceptions.

Critical to all pricing decisions is in-market experience, where price levels can be evaluated and adjusted to reflect such factors as the competitive environment, coupon and display alternatives, and promotional executions.

Product Optimization

At this point, the theoretical product becomes a prototype. R and D may independently or in conjunction with consumer research conduct a product-focused investigation to assess and determine the optimal levels of consumer acceptance in terms of necessary cost constraints and profitability expectations. Research methods at this stage include:

Focus group interviews: Focus group interviews may determine if the product satisfies the attributes requirements previously identified as vital, and may reveal product or positioning issues that must be addressed. Basic areas of concept testing are repeated—benefits, perceived usage, and purchase interest—as well as response to the product, which might include actual product tasting and response to appearance and smell.

In-home usage and evaluation: A number of in-home usage techniques may be employed, which permit consumers to experience the new product's packaging, preparation, and appeal. Respondents may be product-category users, participants in concept development, or persons screened to fit other predetermined criteria. Data then are used to evaluate product fit with the concept, overall consumer acceptance, and future interest in the purchase.

Blind or branded testing: Blind testing evaluates a product without support of any brand name or advertising, and is used to determine whether the test product is equal to or better than a benchmark or competing product. Branded testing evaluates the product's ability to fulfil its concept.

Regression analysis: In regression analysis, another method of product optimization, consumers rate a number of individual product attributes from "excellent" to "poor," in addition to indicating how they like the product itself. Here researchers evaluate how each attribute contributes to the overall rating and determine which attributes emerge as most important to the consumer. If necessary, product reformulation may be undertaken and a second round of testing conducted. Results of the second test will be compared to those of the first to ensure that the product has indeed been improved.

Perception ideal mapping: In perception ideal mapping, consumers rate attributes for both the product being tested and an ideal product. The resulting gap between the product and its ideal measures both the degree to which and areas where the product needs to be improved.

Test Markets and Simulations

Test marketing, whether in simulation or not, is used as a predictive research tool for new product introductions. It uses mathematical models to project forecasts of sales and market share as well as to make recommendations for improvements in pricing, advertising, or promotion.

It is at this point that the product is considered viable and ready to be introduced into an "authentic" setting to judge consumer interest in its purchase and repurchase, in preparation for a national introduction.

Simulated test markets (STMs) were developed in the 1960s to measure, in a controlled setting, trial and repurchase intentions among target users and to estimate year-one volume.

Year-one volume estimates for the new product are made by combining the customer's intent to purchase, the customer's intent to repurchase, and the marketing plan, which factors in such data as advertising, consumer promotions, and anticipated distribution, as well as category purchase cycles and possible competitive pressure.

Minimarket-testing, or controlled test marketing, is conducted by researchers who have arranged for specific stores in specific geographical markets to carry the new product. Display, promotion, and pricing—all facets of the introduction—are strictly controlled and contained. Consumers are not contacted directly before purchase; response to the new product may be evaluated through follow-up interviews later.

A full-fledged test market is a significant investment, involving different stores in a minimum of two locations, and it may take up to a year, thus increasing the likelihood of competitive introductions. However, it affords the new product its first true test of market endurance within the competitive framework.

Qualitative research in the focus group format also may be employed during test marketing to evaluate consumer perceptions of the product and its ability to deliver key benefits, as well as to uncover issues and opportunities for product improvement and the enhancement of consumer satisfaction.

THE NEW PRODUCT LAUNCH

The decision to launch a new product involves marketing research, management judgment and experience, and the commitment of the people and the company involved. Only slightly more than half of all products that go to test market are eventually launched. This is not surprising in view of today's market, where the rate of new product failure and the cost of introduction are high.

The decision to roll out the product on a regional basis only or to introduce it nationally after pretesting may be made on the relative strength of market forecasts, with the strongest results most likely to prompt national introduction. Smaller firms are most likely to begin introduction in specific cities; larger firms are more apt to choose regional roll-outs.

The use of in-market assessments may continue during roll-out, as marketing research focuses on consumer and competitive reaction to the new product and an evaluation of whether product sales can meet forecasts and expectations.

Chapter 2

Market and Consumer Research

An integral part of Food Product Development or any New Product Development is the collection and analysis of background data. The process is what constitutes Market and Consumer Research.

One of the main causes for failure of any new food product is often it being rejected by the consumer. In order to avoid such circumstances, it is important that a thorough background market and consumer research is conducted.

Market research is the study of the needs, patterns, setup and organizational characteristics of the target area or market. It is the overall exhaustive analysis of the market and the marketing strategies to be such that the product is successful and not rejected by the market.

Market Research is "The scientific consolidation of product marketing strategy through the systematic gathering, recording, and analyzing of data about the market and its preferences, opinions, trends, and plans."

It is the systematic collection and evaluation of data regarding customers' preferences for actual and potential products and services.

Consumer research is "The collection, recording and analysis of data pertaining to needs, acceptance policies, food patterns, trends of food consumption, likes and dislikes of the consumers."

The primary goal of consumer market research is to identify, understand, and analyze customers and their needs.

Marketing research focuses on understanding the consumer as a person by focusing on exploring his or her attitudes, needs, motivations, and behaviour as it relates to a product or service **while consumer research** helps to provide a company with relevant, reliable, valid, and current information or their target buyer.

The purposes or objectives of conducting consumer market research are:

o Assisting companies to make better business decisions and gain advantages against the competition from other companies.

o Assisting in the process of making numerous strategic and tactical decisions by executives and/or managers in order to identify and satisfy customer needs.

o Resolving a bit of uncertainty around the marketing variables, environment, and consumers by providing information.

o Providing insights which guide the creation, optimization and expansion concerned with a new product/existing product or service or venturing into new markets.

o Predicting population trends towards likelihood of purchasing a product or service, based on variables such as age, gender, location, and income level.

o Unravelling the trends and pulse of a target market.

o Understanding and identifying customer behaviour, needs and expectations which help in analysis of the product success.

Marketing has many aspects or subdisciplines within the broad discipline of marketing. They include:

o Advertising

o Branding

o Copywriting

o Customer relationship management (CRM)

o Direct marketing

o Event planning

o Graphic design

o Internet marketing

o Loyalty marketing

- Market research
- Marketing communications
- Media relations
- Merchandising
- New product development
- Pricing
- Product management
- Promotion
- Public relations
- Sales management and support
- Search engine optimization (SEO)
- Social media optimization
- Strategic planning
- Supply chain management

Consumer Research—Essential Features

A vital component of consumer research is studying consumer behaviour patterns. Consumer behaviour is the study of when, why, how, and where people do or do not buy a product.

It attempts to understand the buyer decision-making process, both individually and in groups. It studies characteristics of individual consumers such as demographics and behavioural variables in an attempt to understand people's wants. It also tries to assess influences on the consumer from groups, such as family, friends, reference groups, and society in general.

Consumer behaviour study is based on consumer buying behaviour, with the customer playing the three distinct roles of user, payer and buyer.

Factors which dictate consumer behaviour are discussed below:

1. **Psychological factors:** Factors that influence consumer moods and actions are influenced largely by motivation, perception, learning, attitude, personality, and lifestyle. These dictate the consumer spending, buying and using patterns.
2. **Social factors:** Individual actions are to a large extent swayed by social factors like family, social class, reference groups, and culture.

3. **Purchasing power and reason:** These are factors which are related to specific situations which demand spending or buying. Promotion at workplace, mood swings, social reasons, festivals, etc. all greatly influence consumer buying behaviour.

4. **Need based or want based:** An important factor in consumer behaviour is the attitude of the individual, whether the individual is buying due to a need or simply as a wish or want or extravagance. Needs are the basic, motivating forces that shape decision-making while wants are the learned needs that extend beyond the basic needs.

COMPONENTS OF MARKET AND CONSUMER RESEARCH

The first step in the research process is the collection or gathering of data, which is later on converted into information. This is achieved by adopting two possible sources of data collection, viz. primary and secondary.

Primary Sources

These constitute the original or first-hand data. These are the first-hand accounts and are collected keeping in view the intended product itself. It is expensive, and time consuming, but is more focused than secondary research. There are many ways to collect primary data as mentioned below:

1. Interviews
2. Focus groups
3. Projective techniques
4. Product tests
5. Diaries

1. **Interviews:** This is one of the most widely used techniques in primary research and is either interactive or non-interactive. Interactive interviews are telephonic, face-to-face and to some extent via internet while noninteractive is through mail survey. It usually involves the participation of two people an interviewer and an interviewee. The data is collected on the basis of answers to questionnaires.

   a. *Telephone interview:* It is practiced more in developed countries, as most people own a telephone. It is ideal when data has to be collected from a vast geographical area. The interviews tend to be very structured, lack depth and cheaper to conduct than face-to-face interviews (on a per person basis).

b. *Face-to-face interviews:* Here the interviewer is a researcher while the interviewee is the consumer/ respondent. There is a data survey sheet on which the responses are recorded. Some interviews are structured, rigid and use closed questions while some are flexible, in depth, use open questions and are more interactive.

c. *The internet:* This is used to collect primary data when visitors are asked to complete electronic questionnaires.

d. *Mail survey:* Not very common these days yet in some countries it is still the most appropriate way of data collection. The methodology adopted is to collate or purchase lists of respondents contact addresses and a predesigned questionnaire is mailed to them. The response rate varies between 5 and 10% which, in a second mailer or reminder slightly betters. The advent of call centers and telecommunication advancements had made it less popular.

Table 2.1: Comparison table for kinds of interviews

	Telephone interviews	Face-to-face interviews	Internet interviews
Advantages	Can be geographically spread.	They allow more 'depth'.	Relatively inexpensive.
	Can be set up and conducted relatively cheaply.	Physical prompts such as products and pictures can be used.	Uses graphics and visual aids.
	Random samples can be selected	Body language can emphasize responses.	Random samples can be selected.
	Cheaper than face-to-face interviews.	Respondents can be 'observed' at the same time.	Visitors tend to be loyal to particular sites and are willing to give up time to complete the forms.

Contd.

Table 2.1: Comparison table for kinds of Interviews *(Contd.)*

	Telephone interviews	Interviews internet	Face-to-face interviews
Disadvantages	Respondents can simply hang up.	Interviews can be expensive.	Only surveys current, not potential customers.
	Interviews tend to be a lot shorter.	It can take a long period of time to arrange and conduct.	
	Visual aids cannot be used.	Some respondents will give biased responses when face-to-face with a researcher.	Needs knowledge of software to set up questionnaires and methods of processing data.
	Researchers cannot comprehend behavior or body language.		May deter visitors from your website.

Types of questionnaires and questions

The above interviews are possible when one has questionnaires to work on. There are various kinds of questions that constitute questionnaires and any combination can be employed to form questionnaires. The various questions that form questionnaires are as follows:

o *Contingency questions:* These are questions relevant and asked only when the answer to a previous question is in an affirmative. This saves time and energy of the parties involved.

o *Closed ended questions:* The answers to these questions are limited to a fixed set of responses.

- *Yes/No questions:* These questions only have either a positive 'yes' answer or a negative 'no' answer.

- *Multiple choice questions*: The interviewee can choose from several options.

- *Scaled questions:* Responses are graded on a scale (for example, palatability of a food product on a scale from 1 to 10, with 10 being the most preferred appearance).

o *Open-ended questions:* As the name suggests these questions have limitless options as answers, without any boundaries. The respondent forms their own answer without being constrained by a fixed set of possible responses.

- Completely unstructured questions are those that leave the respondent with an open mind and a free will to answer.
- *Word association:* Respondents are presented with some words and they have to say the first word that comes to their mind.
- *Sentence completion:* On being presented with a sentence that is incomplete respondents complete it.
- *Story completion:* On being presented with a story that is incomplete respondents complete it.
- *Picture completion:* On being presented with a picture respondents fill in an empty conversation balloon in it.
- *Thematic apperception test:* Respondents explain a picture or make-up a story about what they think is happening in the picture.

Important points for development of questionnaires are:

o There should be a logical sequence between questions and a natural flow.

o To ensure an unbiased approach the researcher should be careful that an answer should not be dependant on a previous question.

o As a rule, general questions should precede more specific ones and least sensitive questions should always precede (be before) the most sensitive.

o Factual and behavioural questions should be followed by attitudinal and opinion questions.

o According to the sandwich theory, initial questions should be screening followed by product specific questions and lastly demographic questions.

2. **Focus groups:** A selected number of respondents are brought together based on specific criterion. They are called focus groups and they are called in the same room for discussions and opinions to probe into a specific product. A team of expert researchers then interacts with the focus group to gather in-depth qualitative feedback. Generally a focus group would comprise 10 to 18 participants.

Advantages	Disadvantages
Commissioning marketers often observe the group from behind a one-way screen	Highly experienced researchers are needed. They are rare.
Visual aids and tangible products can be circulated and opinions taken	Complex to organize
All participants and the research interact	Can be very expensive in comparison to other methods
Areas of specific interest can be covered in greater depth	

3. **Projective techniques:** They are used to generate highly subjective qualitative data as employed in the field of psychology. Examples of these techniques are Inkblot tests, Sentence or story completion, Word association and Psychodrama.
4. **Product tests:** Sample products are displayed in shopping areas and potential customers are asked to sample them. Thereafter the purchase behaviour is observed by team of observers who then contemplate how the product is handled, packing is read, how much time the consumer spends with the product, and the likes. This data is collected.
5. **Diaries:** As a tool used in monitoring daily behaviour these diaries are given to selected customers who are asked to maintain a diary that lists and records their purchasing behaviour over a period of time (weeks, months, or years).

Secondary Sources

These sources of data collection or research are not first-hand, but already existing records. This exercise is relatively cheaper and quicker. On the flip side since it is not target specific as had been collected for some other purpose the analysis and comparisons might be difficult to make. The sources for such data are numerous like:

- o Trade associations
- o National and local press food magazines
- o National/international governments
- o NGOs
- o Websites
- o Informal contacts
- o Published company accounts
- o Professional institutes and organizations
- o Previously gathered marketing research
- o Public records.

CLASSIFICATION OF MARKET RESEARCH BASED ON TYPES OF RESEARCH DESIGNS

Questioning based		Observation based	
Qualitative research	Quantitative research	Ethnographic studies	Experimental techniques
Generally used for exploratory purposes.	Generally used to draw conclusions.	By nature qualitative, the researcher observes social phenomena in their natural setting.	By nature quantitative, the researcher creates a quasi-artificial environment to try to control spurious factors, then manipulates at least one of the variables.
Small number of respondents.	Tests a specific hypothesis.		
Not genera-lized to the whole population.	Uses random sampling techniques so as to infer from the sample to the population.	Observations can occur: Cross-sectionally (observations made at one time) or longitudinally (observations occur over several time-periods).	

Contd.

Questioning Based		Observation Based	
Qualitative research	*Quantitative research*	*Ethnographic studies*	*Experimental techniques*
Statistical significance and confidence not calculated.	Involves a large number of respondents.	Examples include product-use analysis and computer cookie traces.	Examples include purchase laboratories and test markets.
Examples include focus groups, in-depth interviews, and projective techniques	Examples include surveys and question-naires. Techniques include choice modelling, maximum difference preference scaling, and covariance analysis.		

Researchers often use more than one research design. They may start with secondary research to get background information, then conduct a focus group (qualitative research design) to explore the issues. Finally they might do a full nationwide survey (quantitative research design) in order to devise specific recommendations for the client.

Marketing research from the small and medium-sized enterprises (SME) aspect for start-ups and young entrepreneurs.

Marketing research does not only occur in huge corporations with many employees and a large budget. Small scale surveys and focus groups are low cost ways to gather information from potential and existing customers. Most secondary data can be easily accessed by a small business owner.

Below are some steps that could be done by SME (Small Medium Enterprise) to analyze the market:

1. Provide secondary and/or primary data.
2. Analyze macro and micro economic data.

3. Implement the marketing mix concept: Place, price, product, promotion, people, process, physical evidence and also political and social situation to analyze global market situation.
4. Analyze market trends, growth, market size, market share, market competition.
5. Determine market segment, market target, market forecast and market position.
6. Formulate a strategy and business plan after analyzing the above data.

Overall analysis should be based on "6W+1H" (what, when, where, which, who, why and how) question.

Marketing Research Techniques

Marketing research techniques can take up many forms, including
o **Advertising related:**
  - *Ad tracking*: Periodic or continuous in-market research to monitor a brand's performance using measures such as brand awareness, brand preference, and product usage.
  - *Advertising research*: Used to predict copy testing or track the efficacy of advertisements for any medium, measured by the ad's ability to get attention, communicate the message, build the brand's image, and motivate the consumer to purchase the product or service.
  - *Copy testing*: Predicts in-market performance of an ad before it airs by analyzing audience levels of attention, brand linkage, motivation, entertainment, and communication, as well as breaking down the ad's flow of attention and flow of emotion.
o **Brand related:**
  - *Brand awareness research*: The extent to which consumers can recall or recognise a brand name or product name.
  - *Brand association research:* What do consumers associate with the brand?
  - *Brand attribute research:* What are the key traits that describe the brand promise?

- *Brand name testing:* What do consumers feel about the names of the products?
- *Positioning research:* How does the target market see the brand relative to competitors?—what does the brand stand for?
- *Distribution channel audits:* To assess distributors' and retailers' attitudes toward a product, brand, or company.

o **Consumer related:**
  - *Segmentation research:* To determine the demographic, psychographic, cultural, and behavioural characteristics of potential buyers.
  - *Buyer decision-making process:* To determine what motivates people to buy and what decision-making process they use; over the last decade.
  - Neuromarketing emerged from the convergence of neuroscience and marketing, aiming to understand consumer decision-making process
  - *Commercial eye tracking research:* Examine advertisements, package designs, websites, etc. by analyzing visual behaviour of the consumer.
  - *Concept testing:* To test the acceptance of a concept by target consumers.
  - *Customer satisfaction research:* Quantitative or qualitative studies that yields an understanding of a customer's satisfaction with a transaction.
  - *Mystery consumer or mystery shopping:* An employee or representative of the market research firm anonymously contacts a salesperson and indicates he or she is shopping for a product. The shopper then records the entire experience. This method is often used for quality control or for researching competitors' products.
  - *Internet strategic intelligence:* Searching for customer opinions in the internet: chats, forums, web pages, blogs... where people express freely about their experiences with products, becoming strong opinion formers.
  - *Price elasticity testing:* To determine how sensitive customers are to price changes.

○ **Marketing and sales related:**

- *Demand estimation*: To determine the approximate level of demand for the product.
- *Sales forecasting:* To determine the expected level of sales given the level of demand. With respect to other factors like advertising expenditure, sales promotion, etc.
- *Online panel*: A group of individual who accepted to respond to marketing research online.
- *Store audit:* To measure the sales of a product or product line at a statistically selected store sample in order to determine market share, or to determine whether a retail store provides adequate service.
- *Test marketing:* A small-scale product launch used to determine the likely acceptance of the product when it is introduced into a wider market.

All of these forms of marketing research can be classified as either problem-identification research or as problem-solving research.

THE MARKETING MIX

A famous quote by Sun Tzu goes *"Strategy without tactics is the slowest route to victory. Tactics without strategy is the noise before defeat."*

Therefore there has to be a strategy along with tactics to make a venture successful.

The name given to the strategy adopted based on a few variables to make the product a hit in the market is referred to as the "marketing mix". It is kind of a mixture of parameters that govern whether or not the target audience will accept the product.

In services marketing, a modified and expanded marketing mix is used, typically comprising **seven Ps** made up of the original 4 Ps plus *process, people, physical environment.* Occasionally service marketers will refer to **eight Ps**; comprising the 7 Ps plus **performance**.

In food product development there are also used 5 Ps. They are the variables that marketing managers can control in order to best satisfy customers in the target market. These 5 variables when properly mixed in the best possible ratio would constitute a positive response to the product.

The choice of the 4 Ps, 7 Ps, 8 Ps or 5 Ps is entirely at the discretion of the business owner, entrepreneur or marketing expert based on the goals and objectives of the venture.

The main 5 Ps are:

1. Product
2. People
3. Price
4. Place
5. Promotion

The marketing mix describes how the target market is reached through a specific blend of attributes listed above: **product, pricing, promotion, placement, people** (staff), **process** (of providing a service), **physical evidence** (make service more tangible to potential customer), and **philosophy** (whereby the product reflects the philosophy of the organization).

McCarthy's Four Ps

The original marketing mix, or 4 Ps, as originally proposed by marketer and academic E Jerome McCarthy, provides a framework for marketing decision-making.

Category	Definition/explanation	Typical marketing decisions
Product	A product refers to an item that satisfies the consumer's needs or wants.	• Product design—features, quality
	Products may be tangible (goods) or intangible (services, ideas or experiences).	• Product assortment— product range, product mix, product lines
		• Branding
		• Packaging and labelling
		• Services (complementary service, after-sales service, service level)

Contd.

Category	Definition/explanation	Typical marketing decisions
		• Guarantees and warranties
		• Returns
		• Managing products through the life-cycle
Price	Price refers to the amount a customer pays for a product.	• Price strategy
	Price may also refer to the sacrifice consumers are prepared to make to acquire a product. (e.g. time or effort)	• Price tactics • Price-setting
	Price is the only variable that has implications for revenue.	• Allowances—e.g. rebates for distributors
	Price also includes considerations of customer perceived value.	• Discounts—for customers • Payment terms—credit, payment methods
Promotion	Promotion refers to marketing communications	• Promotional mix—appropriate balance of advertising, PR, direct marketing and sales promotion
	May comprise elements such as: Advertising, PR, direct marketing and sales promotion.	• Message strategy—what is to be communicated • Channel/media strategy—how to reach the target audience • Message frequency—how often to communicate
Distribution (place)	Refers to providing customer access	• Strategies such as intensive distribution, selective distribution, exclusive distribution
	Considers providing convenience for consumer.	• Franchising • Market coverage

Contd.

Category	Definition/explanation	Typical marketing decisions
		• Channel member selection and channel member relationships • Assortment • Location decisions • Inventory • Transport, warehousing and logistics

The 4 Ps have been the cornerstone of the managerial approach to marketing since the 1960s.

MODIFIED AND EXPANDED MARKETING MIX: 7 PS

By the 1980s, a number of theorists were calling for an expanded and modified framework that would be more useful to service marketers. The "seven Ps", which added "physical evidence", "people", and "process" to the original marketing mix, was proposed by Booms and Bitner in 1981.

Category	Definition/explanation	Typical marketing decisions
Physical evidence	The environment in which service occurs. The space where customers and service personnel interact. Tangible commodities (e.g. equipment, furniture) that facilitate service performance. Artifacts that remind customers of a service performance.	• Facilities (e.g. furniture, equipment, access) • Spatial layout (e.g. functionality, efficiency) • Signage (e.g. directional signage, symbols, other signage) • Interior design (e.g. furniture, colour schemes) • Ambient conditions (e.g. noise, air, temperature) • Design of livery (e.g. stationery, brochures, menus, etc.) • Artifacts (e.g. souvenirs, mementos, etc.)

Contd.

Category	Definition/explanation	Typical marketing decisions
People	Human actors who participate in service delivery. Service personnel who represent the company's values to customers. Interactions between customers. Interactions between employees and customers.	• Staff recruitment and training • Uniforms • Scripting • Queuing systems, managing waits • Handling complaints, service failures • Managing social interactions
Process	The procedures, mechanisms and flow of activities by which service is delivered.	• Process design • Blueprinting (i.e. flowcharting) service processes • Standardization vs customization decisions • Diagnosing fail-points, critical incidents and system failures • Monitoring and tracking service performance • Analysis of resource requirements and allocation • Creation and measurement of key performance indicators (KPIs) • Alignment with best practices • Preparation of operations manuals

A modified 5 Ps of marketing mix in FPD has each one of them playing an important role in the entire process.

Marketing is much more than just advertising. Marketing refers to the entire process by which product opportunities are identified, designed, manufactured, and promoted to best fit the needs of consumers.

The various points to consider when developing the marketing mix and deciding on the ratio of the 5 Ps are as follows.

5 Ps MARKETING MIX

5 Ps

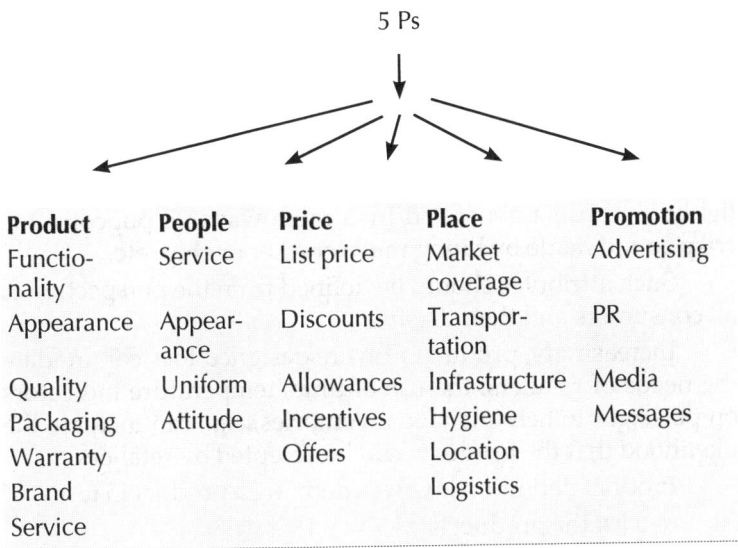

Product	People	Price	Place	Promotion
Functionality	Service	List price	Market coverage	Advertising
Appearance	Appearance	Discounts	Transportation	PR
Quality	Uniform	Allowances	Infrastructure	Media
Packaging	Attitude	Incentives	Hygiene	Messages
Warranty		Offers	Location	
Brand			Logistics	
Service				

Product

The product strategy is complex. It not only includes the development of the correct product attributes to meet the needs of the target market, but must focus on the benefits the customer anticipates the product will provide.

Product refers to a product's attributes as defined by the design objectives mentioned.

All product attributes need to be defined and prioritized in terms of their contribution to the success or failure of the product in the marketplace.

The process of identifying key product attributes may be as basic as defining what consumer needs are to be met by the product, or the inherent advantages that the product offers over its competitors. Or, it may involve quality attributes whose importance is difficult to measure individually in consumer tests, but which collectively have a great impact upon a product's quality perception.

The types of product features consumers desire are:

o Value
o Convenience
o Efficiency in operation
o Dependability in use
o Improvement in earnings

These attributes could include the ease with which a canned fruit cocktail can be opened or an aseptic juice container to spill or splash when opened; the quality and the longevity of the butter aroma generated by a microwavable popcorn; the crisp sound made by biting into a cracker or chip, etc.

Such attributes should be defined from the perspective of all consumers and their lifestyles.

Increasingly, products also are designed to accommodate the needs of retailers; the use of time/temperature indicators on packages to help manage inventories, e.g., can increase the likelihood that the products will be accepted by retailers.

Product definition involves defining a product in terms of

o *what* the product is,
o *who* the consumers are,
o *what* they want.

Product attributes might include flavour, texture, nutritional qualities, appearance, serving size, package material and colour, integrity, size of graphics, and so on; but at some point all product attributes must be prioritized according to the impact they are expected to have upon final product quality and acceptance. The price of omitting key consumer quality variables is great.

Promotion

Product Promotion is *"informing the market about a product, product line, brand, or company and encouraging a purchase decision"*.

The meaning of the term "promotion" extends beyond the act of identifying the ideal promotional venues by which products may be marketed; it also involves the identification of concepts that lend themselves to promotion.

Promotion is a communication between the "seller" (producer, wholesaler, retailer, service provider, etc.) and relevant stakeholders (usually referred to in terms of the Target Market/s). Promotion or marketing communication utilises tools such as advertising, sales promotion, direct marketing, personal selling, marketing public relations and publicity.

Once an organization has learned the market needs, produced or procured a product, and priced it, it then needs to promote the product by letting the market know that it exists, and how it can be purchased.

Promotion involves providing information about a product, product line, brand, or company. There are many ways to promote including:

o Advertising
o Personal selling
o Sales discounts
o Public relations/publicity
o Sampling
o Word of mouth, including electronic endorsements
o Product placement.

Some key points to note in promotion are:

o Products targeted toward children often rely upon shapes and colours like in breakfast cereals, candies, biscuits, etc.
o Product developers have the expertise to know what kinds of shapes can be engineered and which colours are available.
o Sound is a promotable attribute that can be engineered into snacks, like kurkure, nachos, chips, nutty chocolates like crunch, snap, crackle, pop and such sounds, through judicious ingredient and process selection.
o Non-melting properties were designed into Mars M&M® candies and have provided the hook for innumerable promotional campaigns.

In the area of health and nutrition, technical objectives dovetail nicely with promotional objectives. Talking about a specific nutrient in demand will naturally generate consumer interest and skyrocket the sales.

Even nonspecific auras of health and wholesomeness can be promoted. The "natural" foods industry managed to capitalize upon the wholesome connotations of "natural" products even when such value could not be documented.

Similarly, the draw of "all-natural" ingredients provided a technological boon to the flavour industry even though no real social, nutritional, or medical benefits could be attributed to those ingredients.

Price

Price is set at a level which indicates the perceived value agreement between producer and purchaser.

The *price* is set by balancing many factors including supply-and-demand, cost, desired profit, competition, perceived value, and market behavior. Ultimately, the final price is determined by what the market is willing to exchange for the product. Pricing theory can be quite complex because so many factors influence what the purchaser decides is a fair value.

Pricing objectives directly affect an organizations pricing policy:

- **Profit-oriented objectives**: Attain a specific profit level, or as much profit as possible
- **Sales-oriented objectives**: Target a predetermined unit sales level or market share
- **Status quo objectives**: Seek to maintain current price levels, peg them to an index, or match them to a competitor.

Price cannot be separated from the elusive concept of "value." One-way to view value is as the overall impact a product makes upon a consumer's quality of life.

For middle- to upper-income consumers, they may be driven by quality perceptions (e.g. imported wines or cheeses) or convenience (microwavable dinners) for value.

While for some of these products, low price may equate with low value.

For other consumers, value may be dominated by serving sizes, nutritional profiles, and types of packaging.

Lower-income consumers may affix higher values to lower prices.

Too often product developers are constrained from developing products that fully meet consumer's criteria for acceptability because of ingredient, processing, and cost guidelines imposed by agents outside the product development process.

Cost is only one aspect of a product's value and, its importance can often be minimized by other product value attributes.

It should be known that consumers will pay for value— given that a product does, in fact, possess quality.

Placement or Place

Place refers to time and space. **Placement or place refers to the distribution channels that a company will use in order to take its product to the final customer (or consumer).** A product's position on the supermarket shelf is critical.

Delivery of the product to the purchase is product distribution or placement.

At the most basic level, it is necessary to consider where a product rests on the grocery shelf.

o Does it gel well with adjoining products given its appearance?

o Is it the most convenient product to lift, read of pick up on the shelf?

o Is it the most versatile in terms of use, acceptability and storage?

o Does it offer the best price for value?

o Does it have the simplest, most "natural sounding" ingredient statement with respect to its competitors?

Besides this the other aspects are:

o Lifespan

o Timing

o Competition

o Uniqueness.

Once an organization has produced or procured, priced, and promoted a product, it then needs to deliver that product to the purchaser. Some distribution examples:

o Direct sale to the customer from the producer

o Wholesale distribution where the producer sells in large quantities only to an intermediary, not the end user

o Retail sales where a retailer will buy large quantities, but sell smaller quantities to individual customers

o Value added resales (VARs) where an organization purchases a product from a producer and, in turn, resells it to a consumer after adding additional products, services, or expertise.

Chapter 3

Determinants of Food Choices: Trends in Social Change

It is a well-known fact that the three basic needs of humans are food to eat, clothes to wear and shelter to live in. Food is a very important, physiological need of each human and ingestion of food is a simple solution to that need.

Humans are faced with several food choices each day and make decisions on what food to eat based on several criteria. Food preferences are dictated by a number of biological, physiological, psychological, social and emotional factors. Each individual has different food preferences. Food preference can indicate a consumers' choice of one food product over another while liking reflects the assessment of quality of a product.

We all have different learning experiences with food, and this causes different food preferences. These learning experiences are highly individual in nature and can be defined as psychological factors affecting food preferences. Several different factors affect our food choices and preferences; one of these is our biological reactions towards the food we consume.

How we perceive some of the basic tastes, such as our preference for sweet or rejection of bitter tastes may be predetermined. Our chemosensory perception is more or less the same all over the world, and the psychophysical response to sensory qualities may not be much different in different culture. However, the preference for these qualities may depend on the context they are experienced in. Hence, different factors in our surroundings affect our preferences.

Using the Food Choice Questionnaire (FCQ) researchers have revealed that several factors, such as health, price, convenience, mood, sensory appeal, natural content, weight control, familiarity and ethical concerns affect our food choices.

Since people grow-up in different societies all over the world we can see distinct differences in food traditions and cuisines which are dependent on different demographic, socio-cultural and economic factors which modulate the connection between taste responsiveness to food and our choices of food. This entails the assumption that people from the same culture or region of the world would have been affected in the same way culturally and therefore has similar food preferences compared to people from other parts of the world. Further, in the contemporary world it is common for people to eat the same food, or ingredients, all over the world. However, due to differences in weather and soil conditions, it does not necessarily mean that foods are prepared in the same way or that they taste the same across geographical regions.

FACTORS AFFECTING FOOD CHOICES

1. Physiological

Physiological factors are responsible for the need and desire for food by an individual. In order to maintain a healthy body, it should be adequately nourished with food and nutrients. The factors are:

o **Hunger:** The feeling of emptiness, weakness or pain caused by a lack of food is hunger. The intensity keeps getting intense as time passes, until it becomes intolerable. Hunger is controlled by a small gland in the base of the brain called the hypothalamus, which performs a variety of functions like:

  i. controlling body temperature
  ii. regulating appetite, thirst and body fluids
  iii. inducing sleep as well as wakefulness
  iv. controlling the release of growth and sex hormones from various glands throughout the body.

o **Appetite:** The desire for food even when the body is not hungry is termed appetite. It is a reflex, which could be triggered by the sight of appetising food, the aroma of food in preparation, and even the mention of food in conversation. The hypothalamus passed the messages from organs to brain that prompts the salivary glands to stimulate and produce extra fluid causing the mouth to water. Unlike hunger, if appetite is not satisfied it will eventually go away.

o **Nutritional requirements:** It is a choice of many people to choose nutritious food given its health impact on the body. The food ingested should provide essential nutrients that the body can absorb and metabolise. The factors that dictate nutritional requirements are: body type, age, activity level, gender and health status.

- *Body type*: Nutritional requirements, to maintain and operate the body processes, of individuals with larger builds are greater require than more those with a smaller build.

- *Age*: The human body undergoes specific growth stages throughout life right from infancy to old age. Given that nutrients carry-out specific functions within the body, the amount of nutrients needed by an individual is regulated by the growth processes.

- *Activity level*: A sedentary person requires less of all nutrients than an active person. If an individual consumes large quantities of energy but does not move around much, the body stores the excess as adipose tissue (fat).

- *Gender*: Biological activities such as menstruation and childbirth mean that women need to have a higher dietary intake of iron and calcium. Men have a higher proportion of muscle tissue on their bodies, so they require a higher intake of protein than women.

- *Health status*: This determines the kind, type and quantity of food consumed by individuals. Individuals on a specific diet, given a disease conditions, will have modified food preferences than an individual with sound health.

2. Biological Reactions to Food

An individual's personal perceptions of food greatly influence the food choices. Individuals select or reject food based on reaction to the physical appearance, presentation, smell and texture of food. Sensory perceptions refer to the use of senses to gauge the quality and appeal of food and judgments based on these reactions. Some people's food choices are limited because they have a physical reaction to specific foods or ingredients. The reactions vary between individuals but may include abdominal swelling, vomiting, diarrhoea, itching, skin rashes, wheezing, headaches and disturbed sleep.

3. Psychological

Psychological factors relate to the mind and the emotions. These are engrained in an individual and some of them are beliefs, habits, values and past experiences with food.

o **Values:** A value is a deep personal feeling about what is important. Values are strong enough to influence behaviour and motivate action. These may be inherited by family influence or acquired by personal experiences. In terms of food selection, the values most likely to influence choices are related to food origins and the maintenance of health.

- A classic example is vegetarianism, which is often a reflection of value based food selection. A person may find the thought of killing and eating an animal revolting, or they may disagree with the conditions under which some animals are raised as a food source.

o **Beliefs:** Beliefs about what is acceptable to eat vary throughout the world and are often related to religion and cultural heritage. A belief is an opinion or conviction which need not be based on positive scientific proof. Beliefs can be challenged and changed, unlike values that remain fixed. Many religions have food customs and impose restrictions on what their followers eat. Buddhists, e.g. are partial vegetarians. They may eat fish, eggs and dairy products but are not allowed red meat. Some cultural beliefs often prohibit the eating of specific foods and lead to food taboos.

- In an area of mid-Africa, people believe that animal milk is a repulsive body secretion similar to urine. Consequently it is not consumed, despite its nutritional value.

- Some people in remote areas of South East Asia avoid eating eggs or chicken because they are believed to destroy human fertility.
o **Attitudes:** An attitude is the way in which a person views something and behaves towards it, usually after evaluating its merit.
  - Culture
  - Personal history or bad food experiences as a child
  - Travel experiences
  - Perceived status.
o **Habits:** Many of the food choices we make are routine. A habit is something that we do regularly without thinking. Food habits are the same and, like all habits, are difficult to break.
o **Self-image:** Self-image or self-concept is a word used to describe how one feels about oneself. It is the way an individual sees his/her personal appearance including the size, shape and weight of the body.

4. Economic

o **Cost of food:** It is particularly important for low-income families, students and the elderly.
o **Available resources:** Time and money are resources that commonly limit what one buys and prepares for meals. Resources can be interchangeable. Time and money can be used wisely by freezing and safely storing foods that are in season, on sale, or in larger quantities than can be used immediately. The microwave can defrost foods quickly and is cheaper to run than a conventional oven.
o **The marketplace:** The marketplace refers to the place where consumers purchase food. It may be the corner store, the small local supermarket, the suburban shopping centre, or the buying and selling of goods on world markets. Generally, the smaller the selling venue is, the more expensive the food items are.
o **Occupation and finances:** The type of job a person does influence their food selection. The physical demands of the job and its social expectations are reflected in food choices.

5. Social

The cultures or societies that people live in, along with the type of contact that individuals have with one another influence food choices. The type of lifestyle, job and education, size of the family and the importance of hospitality within the social group are also important when making food choices. The way societies and family structures have evolved over the years has changed food choices and preferences to large extent. Some relevant social factors are discussed below:

o **Culture and traditions:** Traditions are customs that are repeated at specific times by members of a group or society. Many traditions relating to special occasions involve food. Festive and social occasions always involve food to some degree, and the meal is often the focus of the event. This influences food choices and preferences that develop over time.

o **Lifestyle:** In general, lifestyle factors that influence food selection relate to:

o **Employment:** The kind of job influences the food choices of an individual. A physically demanding job would require carbohydrate and fat rich foods to sustain the lifestyle while a desk job would require a protein rich diet. Business travellers learn to eat on the go and sample a wide variety of cuisines, picking up new tastes with time.

o **Education:** The more an individual knows about food and nutrition, helps them make wise food choices for themselves and family.

o **Geographic location:** The staple food of a country will most likely depend on whether it can be grown given the geography and climatic conditions. Climate affects not only the types of food grown in an area but also the food choices people make. Summer brings the desire for bright, fresh, light, cold foods while Winter is the season of warm beverages, fried foods, porridge and thick soups.

o **Travel and interests:** Most countries are now open to tourists; the internet allows to make purchases from faraway places; and trade agreements between nations have meant that major events in one part of the world can be felt throughout the rest of the world. All these increase the exposure of individuals to a variety of foods, which influences food choices.

o **Household structures and roles**: The make-up of the family unit determines the variety, quality and quantity of food consumed in a meal. Personal likes and dislikes are often the most important factor in food selection within a household. Catering to different dietary needs within the family may mean that more care needs to be taken to prepare food in an attractive and enjoyable way to capture the attention of all family members like young children, adolescents, adults and elderly.

Social Fabric and Food Choices

Change is life and it is a dynamic process that keeps on taking place. Everything changes and evolves into something new and novel. Novelty is the spice of life and there is no one who would rather live a monotonous life than a vibrant one. Sometimes each one of us craves for change and so either dress differently, go out to new places, make slight changes in our routines or simply change what we normally eat. This is what brings about change. It could be due to our own free will or circumstances, or work pressures, or hectic schedule, or peer pressure or simply to bring ease and comfort in our lives.

Man is a social animal and he is intertwined with society. Each individual influences and brings about a change in society. As a result society has undergone tremendous changes and so have the lifestyles of people. Over the centuries one observes how social changes determine the lifestyle shifts of people. The general ways of life were altered to accommodate social changes.

In historical times when the concept of cooking methods was practically unknown man used to survive on raw food like vegetables, fruits, berries, meat and flesh; all eaten in its natural form, that is raw. Then people lived by hunting wild animals and gathering edible plants. When the herds were plentiful and the plants flourishing, life was good. But, when the herds migrated elsewhere, people had to follow them and often discover a whole new set of plants to supplement their diet. They had a nomadic existence. Later on the nomads started keeping animals in cages and ate them when hungry. Still, they did not know breeding and nurturing of herd. Therefore, the nomads had a feast when they had food and were famished when there was no food.

Eventually some intelligent people worldwide discovered how to preserve meat by drying it. As well as the process by which the seeds of the plants they had been eating and scattered them about, they grew into new plants. Later on with the advent time came agriculture and then began the art of cultivating cereals, grains, pulses and seeds. Domestication of cattle gave milk and dairy products, eggs and meat. This was an evolution parallel with that of human civilizations and gained more and more importance with time. Now food and society are almost intertwined. Thus from practically savage methods of eating we came to sophisticated ways of food consumption. Recent archaeological finds place the beginning of agriculture before 7000 BC and animal domestication (mostly dogs used as hunting aids) thousands of years before that. There is some evidence that the people of Shanidar, in Kurdistan, were domesticating sheep and planting wheat as long ago as 9800 BC. Thereafter the advancements in planting and domestication practices gave rise to settlements and the agriculturist to move from random caves and make shift huts into permanent or semi permanent villages with homes made from stones, wood, or wattle. An early example is the Biblical city of Jericho. It started as such a village around 9000 BC, and has been a settlement of one sort or another ever since. With the advent of time and development of civilizations there came about a more structured and civilized way of life.

Some of the earliest known pottery was found in Catal Huyuk a vast, fertile expanse ideal for primitive agriculture, in Turkey, around 6500 BC. There is also evidence of copper smithing and rope making, and some ovens were big enough to imply that some residents were full time bakers. By 5000 BC, the Euphrates Valley was full of villages and townships.

The initial approach to farming was to remove some of the seeds from food plants before eating them, then scatter the seeds back into the same area they came from. Later, the planters realized that other (non-food) plants were competing with their plants for the field, so they took to weeding the fields to make sure the only their plants were growing there. Everything else was left to nature. Eventually it became obvious that this constant replanting resulted in stunted crops and low yields. The first response was simply to find a new field. After all, the land was vast and people were few. After awhile, though, the obvious fields were used up. Then potential farmers looked to the forests. Thus more and more land came under organised agriculture. Later on people occupied pieces of land and did farming on their pieces. Thus came the concept of ownership which is practiced till today.

EARLIEST TYPES OF COOKING

The origins of cooking are obscure. A hypothesis states that maybe the flesh of a beast killed in a forest fire was eaten by primitive men and found to be more palatable and easier to chew and digest than the customary raw meat.

The revolution came with fire that was accidentally discovered by our ancestors. Man started roasting, boiling, stewing, barbequing, sautéing and frying food.

From whenever it began, however, roasting spitted meats over fires remained virtually the sole culinary technique until the Palaeolithic period, when the Aurignacian people of southern France apparently began to steam their food over hot embers by wrapping it in wet leaves. Until the introduction of pottery during the Neolithic period cooking was restricted to toasting wild grains on flat rocks and using shells, skulls, or hollowed stones to heat liquids.

The earliest compound dish was a crude paste (the prototype of the pulmentum of the Roman legions and the polenta of later Italians) made by mixing water with the cracked kernels of wild grasses. The first crude bread was made when this paste, toasted to crustiness when dropped on a hot stone.

Culinary techniques improved with the introduction of earthenware (and, more or less concomitantly, the development of settled communities), the domestication of livestock, and the cultivation of edible plants. A more dependable supply of foodstuffs, including milk and its derivatives, was now assured. Boiling, stewing, braising, and perhaps even incipient forms of pickling, frying, and oven baking were introduced. Early cooks probably had already learned to preserve meats and fish by smoking, salting, air-drying, or chilling. New utensils made it possible to prepare these foods in new ways.

Now with more changes coming in the social structure it has impacted the food habits as well.

Social Factors affecting Food Choices

Having discussed historical advent of food trends and social set-up we now would concentrate on how the social changes have affected the food trends over the last few decades. The discussions would be done under the following heads:

Family structure: Over the years the family structures have undergone a metamorphosis. From the concept of joint families we have drifted to nuclear and ultranuclear family system. This has resulted in a shift in the eating pattern and food habits to a large extent such that now the traditional meals have disappeared and new fad convenience foods have taken there place. No longer do the "ladies of the house" come together in the kitchen and cook a traditional meal three times a day rather the "lady of the house" cooks food that is less time consuming and easy to make. The sit down dinners have lost their place in the daily lives of people.

Individual lifestyle: Gone are days when women were supposed to stay indoors and typically just take care of the home. These days women walk shoulder to shoulder to the men folk and are progressively taking over the work arena. There are women in every sphere of work and shoulder the responsibility of the home with ease as well. As a result that leaves less time to her disposal and she has to divide time between home and work.

That means less time in the kitchen and therefore she looks for nutritious, easy, convenient and quicker food options. She wants to provide the family good meals with a few constraints as a result of the lifestyle. The work schedules are hectic and so one has to accommodate everything together and strike a balance.

Health aspect: The society these days is fast becoming quite health conscious. The choice of foods as a result has also been influenced. The rage is to eat natural, low calories foods that do not put undue stress on the body and alleviate the population of obesity, as the rates of obesity in the society is increasing. Therefore the focus of food has shifted from the oily, spicy cooked food to the healthier versions. The gym culture is fast catching on and people are exercising more thus eating as per the guidelines of the instructors.

Food variety: The technological and agricultural revolutions have provided access to a large variety and amount of foods to people and so the general common foods have been replaced by trendy novel foods. There consumption is also on the rise and the elite sections of society look for such hard to get foods as a status symbol.

Food quality: The extensive use of pesticides and chemical fertilizers used on our Mother earth has depleted it of its natural qualities and so the food we get today is full of these toxic chemicals which harm the body. In developing counties this is a grave problem while it is not the case on developed nations where there are strict tabs on food quality. Therefore the section of society who can afford it go for imported food-stuffs.

Chapter 4

Food Consumption Trends

Food consumption trends are the eating habits followed by the target group and their behaviour towards various kinds of food. Knowing this becomes important in the science of FPD because this will determine whether the food product will be acceptable by the public or not.

The performance of any economy is measured in terms of the trends and pattern of macroeconomic variable which include national income, consumption, saving, investment and employment. Per capita income and food consumption both are the indicators of human development but food consumption is a better indicator of human welfare.

NEEDS AND TYPES OF FOOD CONSUMPTION TRENDS

The utility of food consumption trends and their need arises once the food product has to be successfully marketed. If a particular ingredient in that food is unacceptable in the target group then obviously no one will buy the product and it will be unsuccessful. Therefore one needs to study the eating habits and pattern in a particular area before developing or launching the product. Products are developed based on the feedback, likes and needs of the people.

Food consumption trends can be used for the following:
1. Monitor the agricultural productivity of a region/country.
2. Analyze the current food situation of a region/country.

3. Make deductions about the food likes and dislikes of a region/country.
4. Access the supply and demand of food commodities in a region/country.
5. Formulate the import/export policies for a region/country.
6. Predict the pattern of consumption of food commodities in a region/country.
7. Evaluate the nutritional status of a given population in a region/country.
8. Design and formulate nutritional intervention programmes in a region/country.
9. Generate data regarding crop production, losses and consumption percentages.
10. Make wise choices regarding new food product development and marketing.

There are various types of food consumption trends and they can be classified on the basis of the following points:

1. **Demographic:** The various demographic parameters that determine the type of food consumption trends are:
   a. age,
   b. gender,
   c. religion,
   d. socio-economic status,
   e. education,
   f. ethnicity,
   g. employment status,
   h. mobility.

   All these factors influence the kind of food people prefer eating and spending on. Some instances of how demographics influence food consumption trends are mentioned below:

   1. Youngsters prefer spicy and global food while the elderly would rather eat low calorie, light and traditional foods.
   2. High income group individuals eat imported foods, processed food and costly foods while the LIG

individuals prefer locally available inexpensive food.

3. Some foods are prohibited as per religion and not consumed by people who follow that religion. For example, beef in Hinduism and pork in Islam.

4. Women generally prefer sour and tart foods like chat, tamarind, raw mango, etc. while generally men do not enjoy these foods that much.

5. Working individuals prefer eating out rather than home, as it is a hassle for them to cook at home everyday.

2. **Food groups**: Staple foods of the people around the world is different as in some parts it is wheat or rice or oats or millets. Some people survive on vegetarian foods while other cannot without meat and meat products. Some refrain from lacto products while others cannot live without them. These various food groups then constitute whether the food pattern is vegetarian or meat oriented, milk or soy based, protein or carbohydrate oriented, oriental or European, traditional or modern, healthy or junk.

Therefore based on the above-mentioned variety of parameters the food consumption trends are also named. These are the types:

a. Wheat consumption trend,

b. Rice consumption trend,

c. Meat based pattern,

d. Vegetable and fruits based pattern,

e. Milk dependant trend,

f. Protein supported pattern,

g. Carbohydrate dependant pattern.

The consumption patterns and trends of food are dependent on a variety of reasons. They could be local or global, regional or national, spiritual or materialistic, therapeutic or indulgent and personal as well.

It has been observed that food consumption of wheat is higher in the North on India as compared to South where rice predominates. Even in northern India the states of Punjab and Haryana have different trends that those practiced in Uttar Pradesh and Bihar.

The question that arises here is what makes it so different and why do people eat as they do?

The answer encompasses various dimensions like:

3. **Sociological dimensions**: The social structure of the area plays a crucial role in the development of food consumption pattern and trends. The kind of people living in that particular area influence what they eat and so a trend evolves. Societies influence these patterns and food is an integral part of it.

Various cultures have multifarious beliefs and food preferences. Most of them are rooted to traditions and rituals. For example, in the North of India wheat harvest is celebrated with so much fan fare and is consumed the maximum. The people there believe that it is the best food as it has historical significance too. South Indians on the other hand take rice as their staple and so consumption of rice there is maximum.

Then festivals are a time to rejoice and each festival in India has special food preparations to adhere to, this also helps in evolving a pattern and sets a trend. Like during the days of Ganesh chaturthi in Maharashtra and related areas the consumption of fruits, vegetables and non-cereal foods increases as people fast.

Different age groups prefer different foodstuffs and the age factor becomes important in determining the patterns. Youngsters prefer more spicy, tasty and global food while the old generation prefers food that is light on the stomach and less spicy.

Besides the atmosphere of the home, family and neighbourhood determine the pattern, as moods of the individual and the environment around determine what and how much they want to eat. All these points constitute the social dimension.

4. **Anthropological dimensions**: Anthropology is the scientific study of the origin, the behaviour, and the physical, social, and cultural development of humans. When it studies the food aspect of it there is focus on

food within a cultural and often cross-cultural context. A subdivision of cultural anthropology is called food-ways and it studies food in the social and cultural setting. The methodology adopted is to study and analyse as well as correlate food remains, the ceramic vessels used to prepare and serve food, and other aspects of the food system with settlement patterns to answer questions about cultural change and the production, storage, distribution, preparation, and serving of food within social contexts.

This can be explained by citing an example of Sissel Johannesseni, an anthropologist, in her study of the people who lived in the central Mississippi River Valley between 500 and 1100 C, combines summaries of internal community patterning, food storage facilities, paleoethnobotanical remains, and ratios of different types of ceramic vessels and sets them within the six-hundred-year period during which this farming society adopted maize agriculture. Her studies explained how they evolved from living as isolated families in the sixth century to group solidarity and affiliation with mound centers. With the development of civilizations there came a shift from household to center, food storage from household pits to above-ground granaries, and greater variation arose in the ways food was cooked and served. Thus knowledge about their food habits could be ascertained. Thus it can be done for today's societies.

5. **Psychological dimension:** The behaviour of the mind is a complex field and comprehension of how and why it reacts in a certain way as it does is very interesting to understand. Food consumption patterns are also governed by the psychological condition of the people. Example is the avoidance of certain foods based on a preconceived notion in the mind that is has an adverse effect on health would be followed as such and no amount of persuasion can change that view. If a person starts to believe that a certain food causes bad luck to them they will stop eating them right away.

Curd-sugar is a good luck charm that is followed in Indian families and students are given a spoon of this when they go for an exam or any important event. Since it is thought to bring good news and success people do it. Even though garlic is a very good therapeutic food some people call it food of the devils and do not eat either garlic or onion.

Besides, a study by the Cornell University Food and Brand Lab infers that people eat more of and report more satisfaction with menu items that have long descriptions instead of simple names. Additionally large package sizes increase consumption by an average 22%, while large movie popcorn buckets led people to eat 45% more even when the popcorn was 10-days-old.

A visual illusion (vertical–horizontal illusion) causes people to pour more liquid into a short wide glass than a tall thin. Thus, the state of the mind and the psychological notions that people harbour greatly influences food consumption pattern and trends.

6. **Economic dimension**: This is a crucial and very important aspect in these times. Economical situation is in a state of fluctuation everywhere there are numerous evolving economies like India and China and some giant plumenting ones. The economy influences the buying power and the standard of living of people. Thus whereas organic foods are all too popular in rich European countries they yet have to become common in Indian set-up. The variety of processed foods is much more in developed world than in developing countries since they will not be successful there. Rising global food prices have also affected the food consumption pattern of people with a shift to low cost food stuffs from, sometimes, the staple food, e.g. rising rice prices have forced some people to shift from this source to cheaper cereals as food and thus a big shift is observed in trends now. If the price of a commodity rises beyond the buying power the people naturally shift to other alternatives thereby causing a change in the food consumption pattern.

FOOD CONSUMPTION TRENDS IN INDIA: IN PERSPECTIVE

India's faster economic growth over 1990s has raised per capita income (expenditure) and has significantly impacted its food consumption patterns by causing a change in the structure of food consumption patterns observed earlier during pre-reforms period.

The consumption pattern in India is defined with the reference to the consumer expenditure survey by the NSSO. These surveys divide rural and urban population into different expenditure group. The distribution of household/person and the per capita monthly expenditure on food and non food items is given for each group.

According to a report published in Food Navigator Asia by William Read in 2013, India's overall food consumption will double by 2030, according to new research by McKinsey and the Confederation of Indian Industry.

The report further says that alongside an expected threefold increase in India's gross domestic product over the same period, food consumption will also rise by 4% each year from Rs 11tn in 2010 to Rs 22.5tn by 2030. To increase the value of agricultural output by 130%, from Rs12.7tn in 2011 to Rs29.3tn in 2030, India must follow 12-point plan to improve yields across all crops, augmenting processing capability and strengthening the quality of farm produce.

"Agriculture needs to get into a mission mode and our analysis shows that instead of focusing on grains and cereals, the first step should be for perishables. Both the Centre and states need to move fast in this sector, to create an enabling policy environment by keeping perishable commodities out of the ambit of the Agriculture Produce Marketing Committee Act," said Adil Zainulbhai, chairman of McKinsey India.

Also, rising affluence will see urban India's food consumption patterns change from being primarily driven by basic foods to more "high-value foods" such as fruits, vegetables and complex proteins.

Currently, India achieves just 50–60% of its potential yield for most crops because of poor technology adoption, weak links between farmers and industry, unexplored opportunities in branding, marketing and exports, lack of infrastructure support and dearth of extension support.

According to the report published by National Council of Applied Economic Research titled "An Analysis of Changing Food Consumption Pattern in India", Indian government is striving to provide food security to all its citizens through various policies and programs. The recently enacted National Food Security Act is the most important one in this direction, which aims to give adequate quantities of cheap cereals (predominantly wheat and rice) to the most vulnerable segment of rural and urban population. Although this effort is laudable, food strategies must not merely be directed at ensuring just food security for all, but must also address providing adequate quantities nutritious, safe and good quality foods which could address the make-up of a healthy diet.

The present study shows that despite rapid economic growth during the past decades, India's average per capita calorie and protein intake has grown only modestly, although the per capita fat consumption has registered a higher growth. Calorie and protein source in the Indian diet is diversifying with fruit/vegetable and animal-based food share increasing and cereal and pulses declining. The implication is that the implementation of the cereal-based National Food Security Act will have only a limited impact in achieving the goal of providing nutritional security to the vulnerable section of the population. There is need to include higher protein food such as pulses or protein-enriched cereals or cereal flours in the program. It is worth mentioning that at present India is exporting a major share of its high protein soybean meal while the country is facing a protein-deficiency. Technology to incorporate soybean products in the diet should be encouraged.

Despite large imports, the overall decline in per capita pulse consumption is also of concern. There is need to increase pulse production in the country as international availability of pulses is limited.

With the rising level of income, per capita fat consumption is growing rapidly and the share of vegetable oil in the overall calorie intake is increasing necessitating large imports. Unless domestic production increases the import requirement will continue to grow with rising per capita income.

India's per capita calorie, protein, and fat consumption remains significantly below that of more developed countries such as China and the United States. The implication is that in coming years with rising per capita income and urbanization, India's demand for various superior food products will continue to increase necessitating a possible change in the food production system and agricultural trade. The implications for the predominant small and marginal farmers could be serious, unless there are incentives and policies that allow them to shift from subsistence agriculture and become more integrated in the global food market.

Deliberations on the potential of the food and agriculture sector to meet the demands and challenges posed by this analysis and its implications for all components in the food chain would be useful.

As indicated in the report "Per capita nutrition supply in India among the lowest in the world", by Dipti Jain, published in the E-paper edition 15 December 2016 in Live mint: India has one of the lowest per capita daily supply of calories, protein and fat, according to Organisation for Economic Co-operation and Development (OECD).

In December 2015, the government promised a 40% increase in the supply of subsidized grains to expand coverage of the food security law. Whether the target of higher food supply coverage will be accomplished is hard to tell, but data show how the supply of nutritious food remains a challenge.

India has one of the lowest per capita daily supplies of calories, protein and fat, according to the Organisation for Economic Cooperation and Development (OECD), a club of rich nations. This is lower than even South Africa, Brazil and Indonesia. Per capita amount of food available is typically calculated as production plus imports minus exports divided by the population.

In India, poor per capita availability is largely a reflection of high poverty, which makes it difficult for a large fraction of the population to access nutritious food. In an October 2015 report, the World Bank said India has been the biggest

WHERE WE STAND

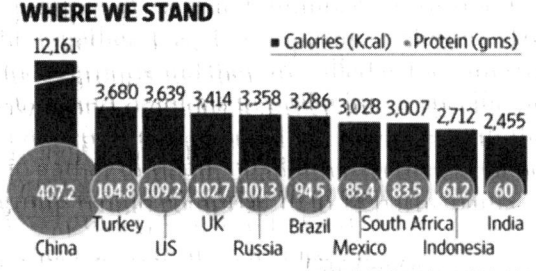

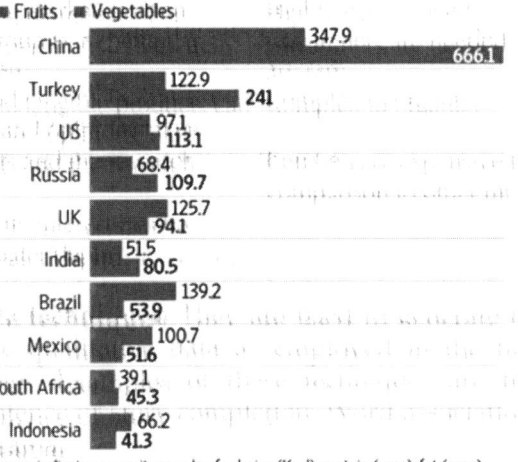

Figures indicate per capita supply of calories (Kcal), protein (gms), fat (gms) per day and sugar, fruits and vegetables (kg per year))   Source: OECD

contributor to poverty reduction between 2008 and 2011, with around 140 million or so lifted out of abject poverty.

"India is a hugely demand-constrained economy. People do not have enough money because of which per capita food supply is low despite high production. If people had enough resources, per capita food availability would have been higher and India would not have the huge buffer stocks it currently has," said Biswajit Dhar, professor at the Centre for Economic Studies and Planning School of Social Sciences at Jawaharlal Nehru University in Delhi.

When demand is low, an increase in local production need not translate into increased availability as a larger portion of the produce may be exported. In India's case, it also depends on changes in government stocks. The Economic Survey show that net cereal production has hardly changed at 465 gm per person per day from 2000 to 2013. However, per capita availability of cereals has increased from 422 gm in 2000 to 468 gm in 2013. However, it was largely a function of changes in government stocks. In the previous two years, availability of cereals was lower at about 410 gm per person per day.

The average Indian had access to 2,455 kcal per day with protein and fat availability at 60 gm and 52.1 gm, respectively, according to OECD. This is far lower than the at least 3,000 kcal

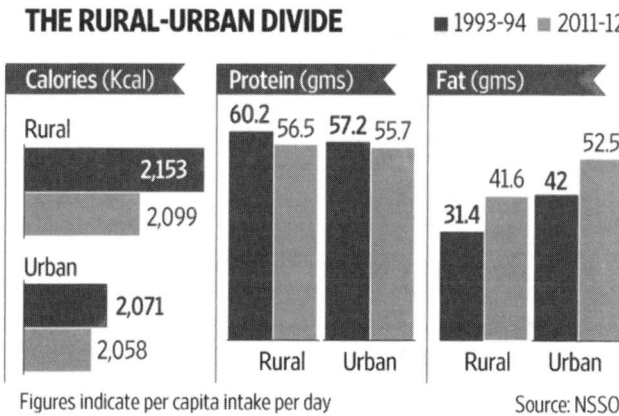

THE RURAL-URBAN DIVIDE ■ 1993-94 ■ 2011-12

Figures indicate per capita intake per day   Source: NSSO

NOT GOOD ENOUGH

	2000	2011
Calories	2,380	2,455
Fat	45.7	52.1
Protein	56.6	60
Sugar	23.3	22.2
Vegetables	64.3	80.5
Fruits	35.5	51.5

(Figures indicate per capita supply of calories (Kcal), protein (gms), fat (gms) per day and sugar, fruits and vegetables (kg per year))

Source: OECD

per day availability for OECD nations. Things have also not improved since the beginning of this century (*see chart 3*). Factors such as wastage of stocks are also to blame for poor availability. For instance, Food Corporation of India data show 3,000 tonnes of foodgrains were damaged in 2015–16. In 2014–15, quantity of damaged grains stood at 19,000 tonnes.

Per capita food availability does not reveal the whole picture. Actual consumption may be even lower depending on the magnitude of losses and wastage in a household, for instance, while cooking and preparation. As data from the National Sample Survey Office show, both total calories and proteins consumed have fallen in the two decades to 2011-12. The average rural Indian consumed only 2,099 kcal per day and urbanites 2,058 kcal per day.

The average intake, according to NSSO, is higher than what is needed. India has a minimum dietary requirement of 1,760 kcal per person per day, according to the Food and Agriculture Organisation (FAO). This number is calculated as the weighted average of minimum energy requirements of different gender and age groups in the population.

However, calorie intake is also clearly a function of income, NSSO figures show. The bottom 5% of the rural population (ranked by per capita expenditure level) consumes only 1,633 kcal per person per day compared with 3,264 for the top 5%.

Poor supply of nutritious food and inadequate per capita income to access good food feeds into health issues such as disability and malnutrition.

"Whether it be poor women's participation in the workforce or wasting and stunting in children under 5 years of age, it can all be addressed to a large extent by supply of good food," Dhar said.

While the government is making efforts to fill shortages in dietary consumption, it is long-term solutions that are truly needed. One fear among experts is that policy makers will start working towards short-term solutions.

"Providing cheap sources of energy in the form of refined cereals, sugars, oils … may lead to earlier onset of risk factors like overweight and obesity, high blood pressure, etc.," said Shweta Khandelwal, Associate Professor at the Public Health

Foundation of India, an autonomous foundation in New Delhi. While calorie and protein consumption have fallen, fat intake has increased both in urban and rural households, NSSO data show.

GLOBAL FOOD CONSUMPTION TRENDS: IN PERSPECTIVE

The primary driving factors for determination of global food consumption patterns are the increase in world population and the increased consumption per capita. The latter is associated with an increase in per capita income.

Since the last decade agricultural and food prices have been on a rise and very volatile. Both developed and developing countries had to face the brunt of higher food prices, and their producers feeling the pressure from higher input costs.

According to the European Union report (2015), over the last 50 years, consumption grew fast across all the major agricultural commodity sectors. The 'golden sixties' remain the period with the highest growth rates for the majority of products. Recently, growth is again accelerating for some commodities, such as oilseeds, cereals and milk, while it is slowing down for other, such as eggs and meat.

The decline in the meat consumption growth in the most recent period is linked to tight supply availability, high prices, the effect of the economic crisis, climate change and possible changes in consumer preferences in developed countries, such as the US and the EU.

Demand in the developed regions has reached a state of maturity, growing at a moderate pace, but demand growth in the developing parts of the world is much faster. The demand epicentre across the world is shifting, with demand from developing countries now exceeding the demand from developed countries. The trajectory in developing countries has been different for animal products, with higher growth rates compared to cereals, but starting from a lower base.

So the world is consuming more, and shifts between commodities occur.

I. Population Growth

One of the most popular drivers cited for demand growth increase is population growth. Although the population in absolute terms is still growing significantly (81.4 million extra in 2013, reaching 7.12 billion), population growth is slowing down, especially in the developed countries. Population is expected to reach the 9.6 billion around 2050 (UN, 2012). Most of the population increase will take place in developing countries.

The main reason for the exponential population growth in the last 50 years has been the decreasing mortality of both infants and elderly, due to better medical care and better nutrition. During the demographic transition, fertility rate also dropped, slowing down the population growth.

Consumption versus Population Growth

World consumption growth outpaces population growth for the major commodity groups, with vegetable oils being the most extreme case. Dairy and meat products show opposite trends, with consumption growth of dairy increasing over the different periods and that of meat decreasing.

Vegetable oil consumption growth is mainly driven by increased use for food, although in more recent periods industrial use and particularly bio-fuel use have gained importance. Palm and soya oils are the main contributors to this high growth.

Cereal consumption is again increasing in recent periods, especially for maize. As opposed to wheat and rice, which have growth rates below or near population growth, maize consumption growth is high and accelerating (at about 4% in the most recent period). This can be linked to the high-yield growth compared to wheat and rice as well as to its multipurpose use for food, feed and fuel production.

World meat consumption growth continues to decline to nearly reach the pace of population growth in the current period. Consumption dynamics differ considerably between developed and developing countries. The split confirms the reduction in meat consumption growth to a level slightly above population growth, both in developed and developing countries. The decrease in developing countries is however more significant, and mainly attributed to beef and poultry. Up to the period 1985–1997, beef and poultry meat consumption was growing fast, with poultry even topping to 9% per year.

In the last two periods, growth rates started decreasing. The beef consumption growth rate has now dropped below the population growth rate, while poultry meat reduced to a level comparable with pig meat. Pig meat consumption showed the highest growth rate in the sixties (nearly 9%) and dropped afterwards, to stabilize in the last period around 3.5%. The current growth reduction can mainly be attributed to the effect of the economic crisis and tight supplies. The shift towards vegetarianism is also a big contributor to reduced meat consumption trends.

Vegetable oils are mainly consumed in the developing world, with high and increasing growth rates. While vegetable oils are mainly used for cooking in developing countries, they are also significantly used as feedstock for bio-fuel production in developed countries. In the 1997–2009 period vegetable oil consumption increased strongly (to 4% per year) due to the lift-off of the biodiesel market. This biodiesel surge seems over now, with the vegetable oil growth rate dropping back to 1% per year, slightly above the population growth rates. Studying the different dairy products shows high and increasing growth rates for butter (to around 3%/year in the last period) and whole milk powder (6%). The growth rate of cheese, the most popular processed dairy product and mainly consumed in developed countries, is considerably above population growth, although it is steadily decreasing over time in developed countries as opposed to developing countries.

II. Changing Income and Consumption per Capita

The increase in population coupled with good employment opportunities indicate that consumption per capita keeps growing. This is linked to the improvement in income per capita across the world, and especially in developing regions. Factors that influence improved income per capita include productivity gains, increased industrialisation, better education, government spending on infrastructure, and higher consumption stimulating the economy.

Net income per capita has steadily increased across the world (by 71% between 1970 and 2012. The BRICS (Brazil,

Russia, India, China and South Africa) experienced a noticeable increase in income per capita since 2000.

III. Diet Pattern Shifts

Consumption per capita has increased substantially over the last decades (both in energy and protein content).

Growth rates are consistently higher in developing countries, but consumption levels per capita are still much lower. Up to today, developing countries have not yet reached the intake levels of the developed countries.

The evolution in diet is mainly influenced by higher income per capita but food prices, individual and socio-cultural preferences, the development of the cold chain as well as other concerns play a role. When focusing on dietary shifts, cereals and sugar have become proportionally less important, as opposed to meat and vegetable oils. The shares of meat, fish and eggs in total protein availability per capita have steadily increased over time. The growth of this share has however stabilised for fish and is negative for meat and eggs in the last period. Dairy products have a stable share of around 10% in total protein intake.

The picture for developed countries contrasts with the one for developing countries, where the diet has diversified compared to the past. Cereals, including rice, as well as vegetable oil, sugar, meat and dairy intake are higher compared to 1961–1973, although in more recent periods cereal intake is stagnating and even declining. The share of cereals also exceeds the share of developed countries. Vegetable oils and all the animal proteins (meat, dairy, fish and eggs) show high and positive growth numbers in the last periods.

Increases in total protein availability per capita are now mainly driven by dairy, fish and pulses. Sugar intake is also stabilising. These numbers seem to suggest that the diet in developing countries is slowly evolving in the direction of the developed countries, with the exception of sugar.

IV. Urbanisation

The world is also becoming increasingly urbanised, and urban population today exceeds rural population. Urban areas typically have a lower relative expenditure on food than rural areas, due to income differences, access to cheap food and different dietary preferences. Rural dwellers tend to eat traditional diets that are high in grains, fruit and vegetables, and low in fat, while groups moving from rural to urban areas experience an increased intake of energy, sugar, refined grains and fats. They also switch to processed foods. Moreover, their diet is richer in animal proteins, with higher consumption of meat, poultry, milk and other dairy products. Urbanisation is therefore considered as a major driving force influencing global demand for animal products, as it stimulates improvements in infrastructure, including cold chains, which allow an increasing trade in perishable goods.

V. Ageing Populations

Due to better income and health care services across the world, world population is also ageing, creating again shifts in dietary habits. As elderly people are more susceptible to diseases, the importance of a balanced energy- and nutrient-rich diet increases.

The analysis of demand showed that the epicentre of increased growth resides in the developing countries. In the coming future it is expected that the growth in general demand will start to slow-down, due to decreasing population growth and stabilising consumption per capita. In the meantime, the dietary pattern in the developing countries will further shift towards the more costly vegetable oils and animal proteins at the expense of cereals.

FOOD CONSUMPTION SCORE

The Food Consumption Score (FCS) is a composite score based on dietary diversity, food frequency, and the relative nutritional importance of different food groups. The FCS is calculated using the frequency of consumption of different food groups consumed by a household during the 7 days before the survey. Scores are clustered into three groups; the results of the analysis categorize each household as having either poor, borderline, or acceptable food consumption.

The Uses of FCS are as follows:
- o Compare food consumption across geography and time.
- o Target households in need of food assistance.
- o Monitor seasonal fluctuations in food consumption.
- o Provide key diet information to early warning analyses.

The Food Consumption Score (FCS), a tool primarily used by WFP, is commonly used as a proxy indicator for access to food.

It is a weighted score based on dietary diversity, food frequency and the nutritional importance of food groups consumed.

Data Collection

Data is collected on the number of days in the last 7 days a household ate specific food items. A seven day recall period is used to make the FCS as precise as possible and reduce recall bias. A questionnaire is prepared and survey done to collect data.

Table 4.1: Questionnaire used to calculate FCS

1. How many days in the last seven days did your household eat....?
   a. write 0 if no consumption of that food item

Food item	Number of days	Food item	Number of days
Rice		Potato (including sweet potato)	
Wheat/other cereals		Dark green vegetables – leafy	
Pulses/beans/ nuts		Other vegetables	
Milk/milk products		Sugar/honey	
Meat		Fruits	
Poultry		Oil	
Eggs		Other food items	
Fish and seafood (fresh/dried)			

Calculating the FCS

The FCS of a household is calculated by multiplying the frequency of foods consumed in the last seven days with the weighting of each food group. The weighting of food groups has been determined by WFP according to the nutrition density of the food group.

Nutrition density is defined as the ratio of nutrient content (in grams) to the total energy content (in kilocalories).

Table 4.2: Detail of food group weights

Food group	Food items belonging to group	Food groups	Weight for FCS
Cereals and grain	Rice, pasta, bread/cake and/or donuts, sorghum, millet, maize	Cereals and tubers	2
Roots	Roots and tubers: potato, yam, cassava, sweet potato, taro and/or other tubers		
Legumes/nut	Beans, cowpeas, peanuts, lentils, nut, soy, pigeon pea and /or other nuts	Pulses	3
Orange vegetables (vegetables rich in vitamin A)	Carrot, red pepper, pumpkin, orange sweet potatoes	Vegetables	1
Green leafy vegetables	Spinach, broccoli, amaranth and/or other dark green leaves, cassava leaves		
Other vegetables	Onion, tomatoes, cucumber, radishes, green beans, peas, lettuce, etc		
Orange fruits (fruits rich in vitamin A)	Mango, papaya, apricot, peach	Fruits	1
Other fruits	Banana, apple, lemon, tangerine		

Contd.

Table 4.2: Detail of food group weights *(Contd.)*

Food group	Food Items belonging to group	Food groups	Weight for FCS
Meat	Goat, beef, chicken, pork (meat in large quantities and not as a condiment)	Meat and fish	4
Liver, kidney, heart and/or other organ meats			
Fish/Shellfish	Fish, including canned tuna, escargot, and/or other seafood (fish in large quantities and not as a condiment)		
Eggs			
Milk and other dairy products	Fresh milk/sour, yogurt, cheese, other dairy products (exclude margarine/ butter or small amounts of milk for tea/coffee)	Milk	4
Oil/fat/butter	Vegetable oil, palm oil, shea butter, margarine, other fats/oil	Oil	0.5
Sugar, or sweet	Sugar, honey, jam, cakes, candy, cookies, pastries, cakes and other sweet (sugary drinks)	Sugar	0.5
Condiments/ spices	Tea, coffee/cocoa, salt, garlic, spices, yeast/ baking powder, lanwin, tomato/ sauce, meat or fish as a condiment, condiments including small amount of milk/ tea coffee	Condi- ments	0

The sum of the scores is then used to determine the FCS. The maximum FCS has a value of 112 which would be achieved if a household ate each food group everyday during the last 7 days.

The total scores are then compared to pre-established thresholds:

- o Poor food consumption: 0 to 21
- o Borderline food consumption: 21.5 to 35
- o Acceptable food consumption: > 35

Limitations

1. Is only a snap-shot of one week food consumption.
2. The FCS does not consider foods consumed outside of the household.
3. It does not provide any information on intra-household food distribution.
4. By collecting data on the number of days each food item was consumed in the last 7 days, it makes it impossible to consider quantity of food eaten.
5. By using a seven day recall period, it provides a short term picture of food security irrespective of seasonality.
6. Does not capture seasonal changes or quantify the food gap.
7. Neither does it capture the intra-household food consumption.
8. Does not show how food consumption has changed as result of crisis, unless previous FCS for same households are available.

• Some people in remote areas of South East Asia avoid eating eggs of chicken because they are believed to destroy sexual fertility.

Attitude: Attitude is the way in which a person views something and behaves towards it, usually after evaluating

Chapter **5**

Traditional Foods: Shifting Trends, Patterns and Innovations

Traditional foods are those foods and dishes which are part of a culture and are passed through generations. It also implies those foods which become part of a culture being consumed over the long-term duration of civilization that have been passed through generations.

Traditional foods and dishes may have a historic precedent in a national dish, regional cuisine or local cuisine. Due to the familiarity and emotional bonding consumers have with traditional foods and beverages, these are made at home regularly or may be produced by restaurants and small manufacturers, and by large food processing plant facilities.

Every region/country has its own traditional cuisine, which is native to that geographical location and is considered exotic in foreign lands.

Talking of the Indian context, the traditional food of India has been widely appreciated for its fantastic blend of herbs and spices. Indian cuisine is known for its large assortment of dishes. The cooking style varies from region to region and within regions. India is quite famous for its diverse multi cuisine available in a large number of restaurants and hotel resorts, which is reminiscent of unity in diversity.

The staple food in India includes wheat, rice and pulses. In modern times Indian pallete has undergone a lot of change. In the last decade, as a result of globalisation, a lot of Indians have travelled to different parts of the world and vice versa, there has been a massive influx of people of different nationalities in

India. This has resulted in Indianisation of various international cuisines. Nowadays, in big metro cities one can find **specialised food joints of international cuisines**.

Some typical cuisines of India are discussed below:

Bengali food: Bengali cuisine is unique in its use of "panchphoron", a term used to refer to the five essential spices, namely mustard, fenugreek seed, cumin seed, aniseed, and black cumin seed. The specialty of Bengali food lies in the perfect blend of sweet and spicy flavours.

Gujarati food: The traditional Gujarati food is primarily vegetarian, with a sweet taste to all dishes and has a high nutritional value. The primary ingredient is Bengal gram flour (besan) in most dishes.

Kashmiri food: Kashmiri food that we have today in the restaurants has evolved over the years. Highly influenced by the traditional food of the Kashmiri pundits, it has now taken some of the features of the cooking style adopted in Central Asia, Persia and Afghanistan.

Mughlai cuisine: Mughlai cuisine is one of the most popular cuisines, whose origin can be traced back to the times of Mughal Empire. Mughlai cuisine consists of the dishes that were prepared in the kitchens of the royal Mughal Emperors. Indian cuisine is predominantly influenced by the cooking style practiced during the Mughal era.

Punjabi food: The cuisine of Punjab has an enormous variety of mouth-watering vegetarian as well as non-vegetarian dishes. The spice content ranges from minimal to pleasant to high. Punjabi food is usually relished by people of all communities. In Punjab, home cooking differs from the restaurant cooking style.

Rajasthani food: The cuisine of Rajasthan is primarily vegetarian and offers a fabulous variety of mouthwatering dishes. The spice content is quite high in comparison to other Indian cuisines, but the food is absolutely scrumptious. Rajasthanis use ghee for cooking most of the dishes. Rajasthani food is well known for its spicy curries and delicious sweets.

South Indian cuisine: The cuisine of South India is known for its light, low calorie appetizing dishes. The traditional food of South India is mainly rice based. The cuisine is famous for its wonderful mixing of rice and lentils to prepare yummy lip smacking dosas, vadas, idlis and uttapams.

BENEFITS OF TRADITIONAL FOODS

Nutritional Benefits of Traditional Foods

Foods from land and sea once provided everything for people. Today, a mix of market and traditional food is common for most people, but traditional food remains an important source of many nutrients. Compared to processed foods and fast foods, traditional foods have an advantage of:

Advantage	Implication
Less calories	Aids in weight control
Less saturated fat	Heart healthy diet
More lean meats and fish	Low in saturated fats and healthier
More iron	Keeps the muscles and blood rich in iron
More zinc	Aids in wound healing and fighting infection
More vitamin A	Enhanced vision and improved infection fighting capability
More calcium	Stronger bones and teeth
Social advantage	Strengthened cultural capacity and well-being

Shifting Trends in Traditional Food Consumption in India

Indian lifestyle has undergone numerous changes and evolved over the years. Each decade marks a shift in the lifestyles and habits of people. Food and taste habits are no exception to this.

Food buying behaviour of consumers in most emerging economies such as India has significantly changed due to an increase in the per capita disposable income, global information and communication technologies, urbanization, education and health awareness and consciousness, movement of households towards higher income groups, changes in lifestyle family structure.

Due to a variety of reasons, traditional foods have been replaced by convenience foods in Indian households. Some are discussed below:

1. **Family structure**
   a. Joint families have been replaced by nuclear families
   b. The percentage of working women have increased rapidly
   c. With both spouses working there is a little time to spend in the kitchen
   d. Due to rise in double income households there is lack of time for cooking traditional breakfast.
   e. Eating out has become a popular option or choice
   f. The desire for convenience food and an increase in the number of working women are some of the important factors driving a strong growth of packaged food products.

2. **Social pressures**
   a. In order to maintain a stature in society, people opt for branded packaged foods
   b. It has become a status symbol as to "who eats out, how often"
   c. Given hectic schedules at work, people have to take out time for social gatherings.
   d. Ordering food/catering is a better option than cooking at home and entertaining

3. **Psychological factors**
   a. Adopting faulty cooking habits at home, with excess fats and sugars, has replaced traditional foods with packaged foods
   b. Being unsure of "how to cook" and no time to learn leaves when more psychologically secure with eating out
   c. Experimenting with exotic and new cuisines is an inherent human nature
   d. Human nature demands a break from the same monotonous traditional foods to trying out new tastes and foods

4. Physiological factors

   a. Increasing cases of obesity, heart diseases and diabetes prompts people to shift from the traditional diet to cornflakes, oats and other health promising foods

   b. Being unaware of the nutritious benefits of traditional foods people resort to cooking with too much oil and sugar resulting in obesity and other diseases

   c. Cereal based food corporates like Kellog's, Quaker, Mohun, Saffola and Britannia are adopting all mechanisms of modern marketing, targeting primarily nuclear households.

5. Economic aspect: Urban/rural divide

   a. Cereals are predominant and cheapest source of calorie and protein for rural masses in the country.

   b. Due to the low level of per capita income, rural masses are not in a position to compensate for nutritional decline due to decline in cereals by increasing consumption of fruits and vegetable, milk, meat etc. to get adequate nutrition.

   c. Prices play an extremely important role in determining food and nutrition security of India's population.

6. Food industry

   a. The food industry has been very successful in replacing fresh and healthy food from consumers' diet with processed food.

   b. Growing sales of processed food, particularly products such as ready meals, can be seen as a key barometer for the westernization of developing markets, and it is therefore no surprise to find that seven of the world's fastest-growing major markets for "meal solution" products are in Asia Pacific.

   c. Food consumption patterns in India are rapidly changing from food products to high value food products and slowly from fresh, unprocessed, unbranded food products to processed, packaged and branded products.

d. With the emergence of the super market and hypermarket culture, consumer preference for packaged food products has increased significantly in the recent years.

Nutrition Transition Model

Gauging the importance of global shifting trends in traditional foods, several hypothesis and explanations have been developed. One of the most talked about is the "Nutrition transition" model which was first proposed in 1993 by Barry Popkin. It refers to the shift in dietary consumption and energy expenditure that coincides with economic, demographic, and epidemiological changes. Specifically the term is used for the transition of developing countries from traditional diets high in cereal and fiber to more western pattern diets high in sugars, fat, and animal-source food.

Popkin used five broad patterns to help summarize the nutrition transition model.

1. The first pattern is that of collecting food, a characterization of hunter-gatherers, whose diets were high in carbohydrates and low in fat, especially saturated fat.

2. The second pattern is defined by famine, a marked scarcity and reduced variation of the food supply.

3. The third pattern is one of receding famine. Fruits, vegetables, and animal protein consumption increases, and starchy staples become less important in the diet.

4. The fourth pattern is one of degenerative diseases onset by a diet high in total fat, cholesterol, sugar, and other refined carbohydrates and low in polyunsaturated fatty acids and fiber. This pattern is often accompanied by an increasingly sedentary lifestyle.

5. The fifth pattern, and most recently emerging pattern, is characterized by a behavioural change reflective of a desire to prevent or delay degenerative diseases.

Recent and rapid changes seen in developing countries from the second and third patterns to the fourth with strong indications to the fifth, is the common focus of nutrition transition research.

This shift is attributable to many causes like:

1. **Globalization** has played a large role in altering the access and availability of foods in formerly undeveloped nations.

2. **Demographic shifts** from rural to urban areas are central to this as well as the liberalization of food markets, global food marketing, and the emergence of transnational food companies in developing countries. All these forces of globalization are creating lifestyle changes that contribute to the nutrition transition.

3. **Technological advancements have made lives easier and less physically taxing.** This causes the altering of energy expenditure that would have helped offset the caloric increases in the diet. Daily tasks and leisure are also affected by technological advancements and contributing to greater rates of inactivity. This increase in calories is due to increased consumption of edible oils, animal-source foods, caloric sweeteners, accompanied by reduced consumption of grains and fruits and vegetables.

4. **Socioeconomic factors** also play an important role as do cultural values tied to appearance and status.

5. **Lifestyle changes**
    i. The forces of globalization are strongly influencing many lifestyle changes in developing countries.
    ii. Daily tasks that were once laborious engagements are now much easier with the help of technological advancements, with examples being appliances such as washing machines, refrigerators, and stoves.
    iii. Recent advancements in the efficiency of food production processes and preservation techniques and improvements in cookware, such as the introduction of improved metal stoves which use fossil fuels and microwave ovens, have helped reduce domestic efforts greatly.

iv. Leisure is being greatly impacted as well. Activities such as playing sports outside are being replaced with television watching and computer games.

v. Decreasing physical leisure activities can also be contributed to urbanization wherein access to fields needed to play such games as football, lawn tennis etc are not available due to such dense populations and their subsequent demand for land.

vi. A major factor is shift in dietary patterns. Diets rich in legumes, other vegetables, and coarse grains are disappearing in all regions and countries. Taking their place are diets characterized by fat-rich edible and vegetable oils, cheap animal-source foods high in fat and protein, and artificially sweetened foods high in sugar and refined carbohydrates.

6. **Biopsychosocial force**

i. Humans have an inherent desire for these new diets and lifestyles from a biological and psychosocial perspective. They have an innate preference for sweets dating back to hunter-gatherer populations. These sweets signaled a good source of energy for hunter-gatherers that were not food secure. It served the purpose of storing food reserves and providing instant energy as well.

ii. Humans also desire to reduce physical exertion. That is why a shift to more sedentary lifestyles from occupational, domestic, and leisurely activities that were previously much more physical taxing.

iii. Socioeconomic and cultural influences also contribute to lifestyle changes associated with nutrition transition. The transfer of tastes by means of tourism and open food trade has introduced developing nations to foods previously enjoyed only by industrialized countries.

iv. Global food advertising and promotion has only further cemented these dietary changes. Additionally some cultures view obese body types in high regard as they relate them to power, beauty and affluence. This is especially true in Mauritania.

Whereas there is a strong tendency, in developed countries, to shift from the fourth to fifth pattern. It is being increasingly seen these days.

An interesting article by Hans Taparia and Pamela Kochnov, highlights this shift in developed countries with special focus to USA.

(*Reference: Hans Taparia, an assistant professor at the New York University Stern School of Business, co-founded and partially owns an organic food business. Pamela Koch is executive director of the Laurie M. Tisch Center for Food, Education, and Policy at Teachers College, Columbia University.*)

"Urbanites are not the only ones turning away from the products created by big food companies. Eating habits are changing across the country and food companies are struggling to keep up.

General Mills will drop all artificial colours and flavours from its cereals. Perdue, Tyson and Foster Farm have begun to limit the use of antibiotics in their chicken. Kraft declared it was dropping artificial dyes from its macaroni and cheese. Hershey's will begin to move away from ingredients such as the emulsifier polyglycerol polyricinoleate to "simple and easy-to-understand ingredients" like "fresh milk from local farms, roasted California almonds, cocoa beans and sugar."

Those announcements reflect a new reality: Consumers are walking away from America's most iconic food brands. Big food manufacturers are reacting by cleaning up their ingredient labels, acquiring healthier brands and coming out with a prodigious array of new products.

Food companies are moving in the right direction, but it would not be enough to save them. If they are to survive changes in eating habits, they need a fundamental shift in their approach.

The food movement over the past couple of decades has substantially altered consumer behaviour and reshaped the competitive landscape.

For the large established food companies, this is having disastrous consequences. Per capita soda sales are down 25 percent since 1998, mostly replaced by water. Orange juice, a drink once seen as an important part of a healthy breakfast, has seen per capita consumption drop 45 percent in the same period. It is now more correctly considered a serious carrier of free sugar, stripped of its natural fibers. Sales of packaged cereals, also heavily sugar-laden, are down over 25 percent since 2000, with yogurt and granola taking their place. Frozen dinner sales are down nearly 12 percent from 2007 to 2013. Sales per outlet at McDonald's have been on a downward spiral for nearly three years, with no end in sight.

To survive, the food industry will need more than its current bag of tricks. There is a consumer shift at play that calls into question the reason packaged foods exist. There was a time when consumers used to walk through every aisle of the grocery store, but today much of their time is being spent in the perimeter of the store with its vast collection of fresh products — raw produce, meats, bakery items and fresh prepared foods.

The outlook for the center of the store is so glum that industry insiders have begun to refer to that space as the morgue. For consumers today, packaged goods conjure up the image of foods stripped of their nutrition and loaded with sugar. Also, decades of deceptive marketing, corporate-sponsored research and government lobbying have left large food companies with brands that are fast becoming liabilities.

For legacy food companies to have any hope of survival, they will have to make bold changes in their core product offerings. Companies will have to drastically cut sugar; process less; go local and organic; use more fruits, vegetables and other whole foods; and develop fresh offerings. General Mills needs to do more than just drop the artificial ingredients from Trix. It needs to drop the sugar substantially, move to 100 percent whole grains, and increase ingredient diversity by expanding to other grains besides corn.

McDonalds needs to do more than use antibiotic-free chicken. The back of the house for its 36,000 restaurants currently looks like a mini-factory serving fried frozen patties and french fries. It needs to look more like a kitchen serving freshly prepared meals with locally sourced vegetables and grains — and it still needs to taste great and be affordable.

These changes would require a complete overhaul of their supply chains, major organizational restructuring and billions of dollars of investment, but these corporations have the resources. It may be their last chance.

Conserving the Traditional Foods: European Perspective

Given the vast shifts from traditional foods to modern foods and an inherent fear that it would cause the complete wipeout of traditional cuisines; there is a growing effort to revive these. This is a movement globally and has followers everywhere.

A commendable step in this direction is an initiative by the European Union.

Some traditional foods have geographical indications and traditional specialities in the European Union designations per European Union schemes of geographical indications and traditional specialties:

Three European Union schemes of geographical indications and traditional specialties, known as **protected designation of origin (PDO)**, **protected geographical indication (PGI)**, and **traditional specialities guaranteed (TSG)**, promote and protect names of quality agricultural products and foodstuffs.

These laws protect the names of wines, cheeses, hams, sausages, seafood, olives, olive oils, beers, Balsamic vinegar and even regional breads, fruits, raw meats and vegetables.

Foods such as Gorgonzola, Parmigiano-Reggiano, Feta, the Waterford Blaas, Herve cheese, Melton Mowbray pork pies, Piave cheese, Asiago cheese, Camembert, Herefordshire Cider, Cognac, Armagnac and Champagne can only be labelled as such if they come from the designated region.

To qualify as *Roquefort*, for example, cheese must be made from milk of a certain breed of sheep, and matured in the natural caves near the town of Roquefort-sur-Soulzon in the Aveyron region of France, where it is colonised by the fungus *Penicillium roqueforti* that grows in these caves.

Protected Designation of Origin (PDO)

The protected designation of origin is the name of an area, a specific place, or in exceptional cases, the name of a country, used as a designation for an agricultural product or a foodstuff o which comes from such an area, place or country,

o whose quality or properties are significantly or exclusively determined by the geographical environment, including natural and human factors,

o whose production, processing and preparation takes place within the determined geographical area.

For PDO status "the entire product must be traditionally and entirely manufactured (prepared, processed *and* produced) within the specific region and thus acquire unique properties."

Protected Geographical Indication (PGI)

Fig. 5.1: Protected geographical indication logo

The protected geographical indication is the name of an area, a specific place, or in exceptional cases, the name of a country, used as a description of an agricultural product or a foodstuff,

o which comes from such an area, place or country,

o which has a specific quality, goodwill or other characteristic property, attributable to its geographical origin,

o at least one of the stages of production, processing or preparation takes place in the area. *"Geographical indications and traditional specialities"*. *European Commission.*

In other words, to receive the PGI status, the entire product must be traditionally and at least partially manufactured (prepared, processed *or* produced) within the specific region and thus acquire unique properties.

Traditional Specialities Guaranteed (TSG)

Fig. 5.2: Traditional specialities guaranteed logo

The TSG quality scheme aims to provide a protection regime for traditional food products of specific character. Differing from PDO and PGI, this quality scheme does not certify that the protected food product has a link to specific geographical area.

For a food name to be registrable under the TSG scheme it must (a) have been traditionally used to refer to the specific product; or (b) identify the traditional character or specific character of the product.

A TSG creates an exclusive right over the registered product name. Accordingly, the registered product name can be used by only those producers who conform to the registered production method and product specifications.

Food Processing

Food is an important part of all cultures. The pleasure of food is enhanced during social interactions. The processing of food from its raw state into a new product contributes considerably to consumers' sensory enjoyment and nourishment, and creates diversity in food preparation and production.

Food processing is the conversion of agricultural products to substances which have particular textural, sensory and nutritional properties using commercially feasible methods. All of the food consumed is processed in some way or the other.

"Food processing is the transformation of raw ingredients, by physical or chemical means into food, or of food into other forms. Food processing combines raw food ingredients to produce marketable food products that can be easily prepared and served by the consumer."

A PEEK IN HISTORY

Food processing dates back to the prehistoric ages when crude processing incorporated fermenting, sun drying, preserving with salt, and various types of cooking (such as roasting, smoking, steaming, and oven baking). Such basic food processing involved chemical enzymatic changes to the basic structure of food in its natural form, as well served to build a barrier against surface microbial activity that caused rapid decay.

Salt-preservation was especially common for foods that constituted warrior and sailors' diets until the introduction of canning methods.

Modern food processing technology developed in the 19th and 20th centuries. In 1809 Nicolas Appert invented a hermetic bottling technique that would preserve food for French troops which ultimately contributed to the development of tinning, and subsequently canning by Peter Durand in 1810. Pasteurization, discovered by Louis Pasteur in 1864, improved the quality of preserved foods and introduced the wine, beer, and milk preservation.

In the 20th and 21st centuries the space race and the rising consumer society in developed countries, contributed to the growth of food processing with such advances as spray drying, evaporation, juice concentrates, freeze drying and the introduction of artificial sweeteners, colouring agents, and such preservatives as sodium benzoate and food products such as dried instant soups, reconstituted fruits and juices, and self-cooking meals.

Foods are processed for five major reasons:

o Preservation for later consumption or sale to fetch better price

o Removal of inedible portions

o Destruction or removal of harmful substances

o Conversion to forms desired by the consumer and

o Subdivision into food ingredients.

Foods such as fruits, nuts and some vegetables can be consumed in their raw state. However, most food undergoes various degrees of processing prior to appearing at the table. The whole process is sometimes referred to as 'paddock to plate' because it broadly describes the stages involved in production of food from its source of origin to the consumer.

Food processing procedures aim at altering the properties of food, viz. sensory, physical, chemical and/or functional properties of foods.

Some of these are discussed below.

The Chemical Properties of Foods

The chemical substances or compounds contained in foods are nutrients, enzymes, pigments and flavours. These substances are the natural components of food and form its chemical properties. The amount and type of chemicals influence the physical and sensory properties along with the nutritional value of food. The amount, type and activity of the chemical properties cause changes to be observed during growth, harvesting, fishing and slaughtering processes.

Further changes to the chemical properties that occur throughout food preparation, cooking and preserving can be desirable, though some are undesirable. The natural components perform a variety of functions, contributing to the success of food products throughout preparation, cooking and preservation activities.

The hanging of a meat carcass after slaughtering allows a natural enzyme action to occur, which produces more tenderness in the meat. Some of the natural components of a food can be exploited to make a new product from the fresh ingredient, such as the pectin contained in fruits when combined with fruit acids and sugar, which is essential in setting and preserving jams and jellies.

The presence of lactic-acid-producing bacteria, a natural component in unpasteurised milk, is important in the production of some cheeses and fermented milk products.

The activity of the natural components is influenced by other conditions: The food being acidic or alkaline (indicated by its pH level), the temperature of the food or the cooking medium, and other circumstances such as the application of heat or the use of mechanical action, and the presence of oxygen.

The Functional Properties of Foods

Many chemical components have functional roles to perform in food preparation. Functional roles of the natural components of food are referred to as the functional properties and are the results of the combination of the chemical and physical properties that are utilised in food preparation and processing.

The functional properties of food are described as the impact that the physical and chemical properties have on the outcomes of food preparation and food processing.

The process is categorised according to the range of activities that are performed on the food from where it is sourced to where the consumer obtains the food. Commercial food processing is conducted in two broad stages: Primary processing and secondary processing.

Another processing category is called tertiary processing.

A range of activities are performed at each stage. Commercial processing includes activities such as transportation, sorting, cleaning, blending, cooking, preserving, packing, marketing and storage.

Most of the foods available to consumers undergo primary processing, commencing at the point of origin and including the transporting, cleaning and sorting of the raw food.

Some foods undergo secondary processing, in which primary products are changed into other types of products.

- o Wheat can be transformed from the original grain to food products such as flour, breads and cakes.
- o Milk can be converted to cheese and yogurt.
- o Fruit can be processed into jams, jellies and marmalades.
- o Meats can be used as components of frozen meals or preserved as processed meats such as salami and ham.

Primary Processing

Primary processing is the conversion of raw food materials to foods that can be eaten or to ingredients that are used to make edible food products. At its simplest it can be seen as: Washing, milling, trimming, squeezing, peeling, ageing and butchery, shelling and chopping. However, complex processes used in the production of vegetable oils, sugar, wine, milk, tea and coffee are also seen as 'primary' processing: Extraction and refining, pasteurisation and fermentation.

Reasons for Primary Processing

- o Prepare raw food so that is ready for human consumption.
  - Raw food is prepared by cleaning and sorting so that the produce is available to the consumer in an appealing and useable form.
  - Removing insects, soil, stones, and other debris from the edible parts of the food extends the shelf life of the food and improves its immediate sensory appeal to the consumer at the marketplace.

○ Make raw food available to the consumer despite their geographical location and the season of the year.

○ Extend the shelf life of the raw food by placing it in suitable conditions so that it is still in peak condition when transported to a food manufacturer or consumer marketplace.

- Tomatoes are refrigerated in bulk after harvesting so that they will not spoil prior to transportation to a food manufacturer or marketplace.

- Potatoes and apples are stored in cold storage facilities for availability year round.

- Winter vegetables are blanched and dried for availability during summer months.

○ Test raw food for quality assurance.

- Manufacturers seek to have a product that has a consistent and reliable quality. Harvested wheat grain is stored in silos and tested for moisture content so that the grain cannot spoil during storage. Moisture content is also an important factor when selling the grain to food manufacturers.

- Hanging meat carcasses is essential to allow maximum tenderness of the meat for the food manufacturer or consumer.

- The nutrient content of milk is tested for quality assurance and to monitor the yield and health of the herd.

○ Protect food from contamination.

- The temperature in a wheat silo is constant and the air is replaced with a non-toxic gas that kills rodents and insects.

○ Prepare raw food for delivery to food manufacturers for conversion into other food products.

- Wheat is milled and converted into flour. The flour can be transported to a food manufacturer for blending with other ingredients in the manufacture of bread, pasta, noodles, soups and sauces.

Secondary Processing

Secondary processing is the conversion of raw ingredients, i.e. the products of primary processing, to edible food products.

This can occur by changing the product's chemical and physical properties, such as producing skim milk from full-cream milk, or by combining ingredients that alter properties to create a food such as cheese, or by combining many different ingredients to create a vastly different product such as a strawberry cheesecake.

These procedures may involve one or more of the following: Mixing, heating, enrobing, cooling, extruding, drying, layering/dividing, aerating, forming/moulding and fortifying.

Baking cake is an example of secondary processing. A considerable variety of products can be made using similar basic ingredients, e.g. to make bread we need flour, water and salt. Breads such as chappatti and pita are examples of unleavened breads. If yeast is added this produces carbon-dioxide, given the correct conditions, this raises the mixture to produce breads.

Reasons for Secondary Processing

- Converting raw foods into edible food products make them palatable and convenient for consumers to use as fresh produce, such as baked goods from a bakery and preserved products such as canned, bottled, frozen and dehydrated food items.
- Secondary processing standardises food quality.
  - Foods are processed soon after they are harvested, while at their peak.
- People live long distances from where most food is produced.
  - To prevent too much food being spoiled through transporting it to urban areas, secondary processing largely occurs close to where food is grown.
- The amount of food required increases as population increases. By processing food it can be stored for use when required.
  - Many perishable foods are seasonal; that is, they can be grown and harvested only at a certain time each year. By processing foods from an annual harvest, they can be made available to consumers throughout the whole year.

○ Secondary processing increases choice and provides greater variety in meals.
  • Many fruits and vegetables, which in the past would have been available for only short periods during the year, are now available all year round. This allows consumers increased choices in the selection of food from a greater range.
○ The range of processed foods available enables consumers to prepare easy and convenient meals, saving preparation time and simplifying food preparation in the home.
○ Foods are also prepared and packaged for many different storage situations.
  • Packaging is made so that it is appropriate for the food and the storage method.
○ Processing foods destroys many micro-organisms, which can cause disease, making them safer for consumption.
○ Just after harvest, food is at its prime and contains high levels of nutrients that may be lost during processing.
  • Nutrients may be added to processed foods during secondary processing to replace those lost during primary processing or to enrich the product.
○ Transport and marketing of foods influence processing.
  • Time is needed to get foods to the point of purchase.
  • Most perishable foods require some processing to ensure they have an adequate shelf life.
  • An adequate shelf life allows foods to be stored in the home after purchase. Secondary processing helps to increase a food's shelf life.
○ Secondary processing can add value to an existing food or food product.

Primary processing	Secondary processing
The processing that occurs after harvesting or slaughter to make food ready for consumption or use in other food products	Turns primary processed food into other food products

Contd.

Advantages	Advantages
• Easily transported	• Can be used for a number of purposes
• Ready to be sold	• Do not spoil quickly
• Ready to be eaten	• Are available all year (e.g. seasonal foods)
• Ready to be processed into other products	

- Salad combinations sold in supermarkets include pre-pared vegetables and other ingredients such as salad dressing and herbs. This gives added value to the original products of the salad vegetables.

Tertiary Processing

Convenience food, or tertiary processed food, is commercially prepared food designed for ease of consumption. Products designated as convenience foods are often pre-prepared food stuffs that can be sold as hot, ready-to-eat dishes; as room temperature, shelf-stable products; or as refrigerated or frozen products that require minimal preparation, typically just heating, by the consumer. These products often are sold in portion controlled, single serve packaging designed for portability for "on-the-go" or later eating. Convenience food can include products such as candy; beverages such as soft drinks, juices and milk; fast food; nuts, fruits and vegetables in fresh or preserved states; processed meats and cheeses; and canned products such as soups and pasta dishes.

The Pros and Cons of Food Processing

The advantages of food processing are:
o Health aspect
- Makes many kinds of foods safe to eat by de-activating spoilage and pathogenic micro-organisms
- Modern food processing also improves the quality of life for people with allergies, diabetics, and other people who cannot consume some common food elements.
- Processed foods are usually less susceptible to early spoilage than fresh foods.
- Toxin removal

- Lowering of moisture, subjecting to high temperatures, etc. thereby reducing microbial contamination.
- Processing can also reduce the incidence of food borne disease. Fresh materials, such as fresh produce and raw meats, are more likely to harbour pathogenic micro-organisms (e.g. Salmonella) capable of causing serious illnesses.
○ Sensory qualities
  - The act of processing can often improve the taste of food significantly
○ Shelf life
  - Processing techniques help to enhance the shelf life of products by minimizing favourable conditions for microbial growth like moisture, pH and temperature.
○ Economic aspect
  - Mass production of food is much cheaper overall than individual production of meals from raw ingredients.
○ Nutritional aspect
  - Some processed foods help to improve the overall nutrition of populations as it makes many new foods available to the masses.
  - Food processing can also add extra nutrients in the foods through fortification and enrichment processes.
○ Food security and accessibility
  - Increases yearly availability of many foods
  - Some processed foods help to alleviate food shortages.
○ Convenience factor
  - In many families the adults are working away from home and therefore there is a little time for the preparation of food based on fresh ingredients.
  - The food industry offers products that fulfill many different needs, e.g. fully prepared ready meals that can be heated up in the microwave oven within a few minutes.
  - The increase in free time allows people much more choice in lifestyle than previously allowed.
○ Transportation
  - Enables transportation of delicate perishable foods across long distances

o Increasing exposure to varied foods
  • Transportation of more exotic foods, as well as the elimination of much hard labour gives the modern eater easy access to a wide variety of food unimaginable to their ancestors.

The disadvantages of food processing are:
o Nutritional aspect
  • Any processing of food can affect its nutritional density.
  • The amount of nutrients lost depends on the food and processing method.
    ▪ An example is that heat used during processing destroys vitamin C. As a result, canned fruits possess less vitamin C than their fresh alternatives.
o Microbial gut flora
  • New research highlights the importance of a rich microbial environment in the intestine to human health. While it indicates that abundant food processing (with the exception of fermentation) endangers that beneficial microbial environment.
o Use of food additive
  • Using food additives represents another safety concern
  • The health risks of any given additive vary greatly from person to person.
    ▪ sugar as an additive endangers diabetics while salt endangers hypertension patients.
    ▪ As effects of chemical additives are learnt, changes to laws and regulatory practices are made to make such processed foods safer.
o Food contamination risks
  • Various points of microbial/ dust contamination due to many mechanical processes involved that utilize large mixing, grinding, chopping and emulsifying equipment in the production process.
  • Metal contamination risks
    ▪ Overtime and overuse causes machine parts to rupture or fracture.
    ▪ Due to this the product stream gets small to large metal contaminants.

- Further processing of these metal fragments will result in downstream equipment failure and the risk of ingestion by the consumer.

Note: As a precautionary measure, food manufacturers utilize industrial metal detectors to detect and reject automatically any metal fragment. Large food processors will utilize many metal detectors within the processing stream to reduce damage to processing machinery as well as risk to consumer health.

Table 6.1: Typical maximum nutrient losses due to cooking

Vitamins and minerals	Freeze	Dry	Cook	Cook + drain	Reheat
Calcium	5%	0%	20%	25%	0%
Copper	10%	0%	40%	45%	0%
Folate	5%	50%	70%	75%	30%
Folic acid	5%	50%	70%	75%	30%
Food folate	5%	50%	70%	75%	30%
Iron	0%	0%	35%	40%	0%
Magnesium	0%	0%	25%	40%	0%
Niacin	0%	10%	40%	55%	5%
Phosphorus	0%	0%	25%	35%	0%
Potassium	10%	0%	30%	70%	0%
Riboflavin	0%	10%	25%	45%	5%
Sodium	0%	0%	25%	55%	0%
Thiamin	5%	30%	55%	70%	40%
Vit A: Alpha carotene	5%	50%	25%	35%	10%
Vit A: Beta carotene	5%	50%	25%	35%	10%
Vit A: Beta cryptoxanthin	5%	50%	25%	35%	10%
Vit A: Lutein + zeaxanthin	5%	50%	25%	35%	10%
Vit A: Lycopene	5%	50%	25%	35%	10%
Vit A: Retinol Activity Equivalent	5%	50%	25%	35%	10%

Contd.

Table 6.1: Typical maximum nutrient losses due to cooking *(Contd.)*

Vitamins and minerals	Freeze	Dry	Cook	Cook + drain	Reheat
Vitamin A	5%	50%	25%	35%	10%
Vitamin B$_{12}$	0%	0%	45%	50%	45%
Vitamin B$_6$	0%	10%	50%	65%	45%
Vitamin C	30%	80%	50%	75%	50%
Zinc	0%	0%	25%	25%	0%

(Reference: "USDA Table of nutrient retention factors, release 6" USDA. Dec 2007)

MEASURING EFFECTIVENESS OF FOOD PROCESSING PROCEDURES

It is extremely important to measure the effectiveness and performance of the food processing procedures. There are a few parameters that help in doing so. Some of them are discussed below.

o **Hygiene**—measured by how free of contamination are the products produced/processed, e.g. number of micro-organisms per mL of finished product.

o **Energy efficiency measured**—measured by how much product is produced/processed using how much energy, e.g. "ton of steam per ton of sugar produced".

o **Minimization of waste**—measured by how much waste is generated during production/processing, e.g. "percentage of peeling loss during the peeling of potatoes".

o **Labour used during the process**—measured by labour used for production/processing of a specified amount of product, e.g., "number of working hours per ton of finished product".

Standards and considerations employed in modern food processing

o **Health**

• Reduction of fat content in final product by using baking or air-frying instead of deep-frying processes.

- Maintaining the natural taste of the product by using less artificial sweetener than was used before.
- **Hygiene**
  - The rigorous application of industry and government endorsed standards to minimise possible risk and hazards.
  - The international standard adopted is HACCP.
- **Efficiency**
  - Rising energy costs lead to increasing usage of energy-saving technologies.
    - frequency converters on electrical drives
    - heat insulation of factory buildings and heated vessels
    - energy recovery systems
  - Factory automation systems reduce personnel costs and may lead to more stable production results.

Food Processing and Agriculture: A Current Indian Scenario

Indian agriculture is a way of life and it supports about 60 per cent of population for their livelihood and contributes 17% of GDP in India. Engineering inputs are vital for modernization of agriculture, agro-processing and rural living. It is needed for development and optimal utilization of natural resources, appropriate mechanism of unit operations of agriculture for increasing production, productivity with reduced unit cost of production for greater profitability, economic competitiveness and sustainability.

Need for Food Processing

Agricultural produce and by-products are perishable in nature in varying degree and their perishable nature gets exploited on the market floor compelling distress sales orchestrated by factors of demand and supply, intervention of the faces of marketing in the absence of matching post-harvest technology (PHT) and agro-processing infrastructure.

Emerging Trends and Constraints in Food Processing Sector

To meet the current demand of food materials, the industrial food processing sector has emerged. About, 42% of the output comes from the unorganized sector, 25% comes from the organized sector and the rest of it comes from the small-scale players. The small-scale food processing sector is a major source of employment and adds value to crops by processing. It is a major source of food in the human diet.

○ The small-scale food processing sector is, however, under increasing threat and competition from the large manufacturers, who through economies of scale and better presentation and marketing.

○ Good packaging lies at the very heart of presentation and thus customer appeal. It is an area of vital importance for small and medium food manufacturers if they are going to continue to compete and expand.

○ With food processing, it is possible to maintain a nutritious and safe food supply for the millions of people that inhabit both urban and rural areas.

  • Improvement in processing efficiency, by increased yield of usable product, is a tangible means of reducing food loss and increasing food supply.

○ Demand for increased convenience of food preparation in the home, institution and restaurant has created a need from processing industries for food ingredients as well as new food forms.

○ Field crops, including grains, oilseeds, sugar crops and forages are major contributors of the nutrients required by man either through direct consumption of the seed kernel or isolated components as food, or through utilization of the plant and byproducts as feed in the production of meat, poultry, milk, eggs and fish.

  • In essentially all instances, harvested field crops must be processed in some manner prior to utilization as food or feed or in industry so as to reduce their post harvest losses.

India is the world's second largest producer of food next to China, and has the potential of being the biggest with the food and agricultural sector. The total food production in India is likely to double in the next ten years and there is an opportunity for large investments in food and food processing technologies, skills and equipment, especially in areas of canning, dairy and food processing, packaging, frozen food/ refrigeration and thermo-processing. Fruits and vegetables processing, fisheries, milk and milk products, meat amd poultry, packaged/convenience foods, alcoholic beverages and soft drinks and grain processing are important sub-sectors of the food processing industry. The consumer product groups like confectionery, chocolates and cocoa products, soya-based products, mineral water, high protein foods, soft beverages, alcoholic and non-alcoholic fruit beverages, etc. along with the health food and health food supplements is another rapidly rising segment of this industry which is gaining vast popularity.

India produces nearly 16% of the world's total food grain production. It is one of the largest producers of agricultural produce. With a population expected to reach to about 590 million people by 2030 in urban India, India has a huge potential domestic demand for processed foods other than the demand from the exports. There are many socio-economic factors that are driving the demand side of the Indian Food Processing Industry.

The changing consumption patterns, both in tier 1 and tier 2 cities, rising income levels among the middle-class and changing lifestyles, are some of the factors providing the demand side push for the food processing industry. Moreover, the central government has given a priority status to all agro-processing businesses.

Limiting Factors in Food Processing Industry in India

1. **Capital intensive**
    a. It creates a strong entry barrier and allows lesser number of players to enter the market.
    b. Lesser players mean lesser competition and lesser competition means reduced efforts to improve the quality standards.
2. **Poor infrastructure for storing raw food materials**
    a. Two main types of storages—the warehouses and the cold storages, lag in storage standards.
    b. The pests infest the grains sometimes due to lack of monitoring, proper use of pesticides and proper ventilation.

c. Power outages result in sub-optimal function of the cold storages and the quality of food material in the cold storages becomes questionable.

3. **Poor quality standards and control methods** for implementing the quality standards for processing and packaging the processed foods.

4. **Absence of**
   a. Continuity of quality power
   b. Good quality of water for processing
   c. Instruments for rapid and reliable analysis, versatile instruments/equipment for multi commodity.

Chapter 7

Convenience Foods

Technological developments in the field of food processing equipment, processes and packing material have brought revolution in the development of convenience foods as per the necessity, taste as well as nutritional requirements of the consumers.

Convenience foods or **tertiary processed foods** are commercially prepared foods designed for ease of consumption. Convenience foods are those foods to which some or all of the labor for preparation has been added at the time of purchase. Convenience foods can be defined as foods that have undergone major processing by the manufacturer such that they require a little or no secondary processing and cooking before consumption.

Products designated as convenience foods are often prepared foodstuffs that can be sold as hot, ready to eat dishes at room temperature, shelf stable products or as refrigerated or frozen products that require minimal preparation typically by just heating.

It may also be easily portable, have a long shelf life, or offer a combination of such convenient traits. Convenience foods have also been described as foods that have been created to "make them more appealing to the consumer." Although restaurant meals meet this definition, the term is seldom applied to them. They differ in that restaurant food is ready to eat, whilst convenience food usually requires rudimentary preparation. Both typically cost more money and less time compared to home cooking from scratch. Convenience foods and restaurants are similar in that they save time.

A food may be classified as convenience food if it meets the criteria like:

o The food must have undergone considerable amount of food preparation by manufacturer before it reaches the retailer.
o It has a longer shelf life and is easily portable.
o It must require minimal cooking or processing before consumption by the consumer.
o The preparation time before consumption should be minimal.

Evolution, Innovation or Introduction of Convenience Foods

Although it would be impossible to name one single date when such foods were first invented or the term first coined, it is evident that convenience foods or ready-to-eat foods began as simple single ingredient snacks.

The evolution of convenience foods:

o Throughout history, people have bought food from bakeries, creameries, butcher shops and other commercial processors to save time and effort.
o The frequency increased with time and people started using preservation techniques to save time and energy in food preparation. Dried vegetables, dried cake mixes, meal mixes, powdered spice mixes all were developed.
o It took a step further with canned food, which was developed in the 19th century, primarily for military use, and became more popular during World War I.
o One of the earliest industrial-scale processed foods was meatpacking. With the advent of a system of refrigerator, cars in 1878 in USA, meat could be raised, slaughtered, and butchered hundreds (later thousands) of miles or kilometers away from the consumer.
o Experience in World War II contributed to the development of frozen foods and the frozen food industry.
o Modern convenience food saw its beginnings in the United States during the period that began after World War II.

- o Thereafter there was no looking back and companies dived in deep in this sector resulting in a flood of such products in the markets globally.
- o As of the 2010s due to increased preference for fresh, "natural", whole, and organic food and health concerns the acceptability of processed food to consumers in the developed world is dropping and the reputation of major packaged food brands had been damaged.
- o Firms responded by offering "healthier" formulations and acquisition of brands with better reputations.

There are also several interesting anecdotes that detail how shifts from conventional cooking to prepared snack foods may have first originated.

The Invention of the Ever-popular Potato Chip in 1853

A cook by the name of George Crum at an upscale resort in Saratoga Springs, New York is attributed to this. A patron there had ordered some fried potatoes with his meal, which was already a common menu item in that region. On being served, he complained that the potatoes were too thick and sent them back to the cook. Crum, who was known to possess a fiery disposition, was rather upset that someone had criticized his cooking. So he sliced a new batch of potatoes paper-thin, fried them crisp in boiling oil, and then salted them.

Very soon the chips became famous and were termed Saratoga Chips. George Crum later started his own restaurant called Crumbs House, and promoted the chips further by placing large baskets of these on every table. They soon became popular all over and came to be known as potato chips. Commercial manufacturing followed shortly, and they became available in grocery stores in 1895.

Various Aspects of Convenience Foods Tabulated

Advantages	Disadvantages
• Preparation time is reduced to a great extent.	• Cooking time is sometimes increased for thawing or longer baking time.

Contd.

Advantages	Disadvantages
• No storing, buying or planning of ingredients.	• Harder to control fat, salt and sugar levels.
• Due to calculated portion sizes, hardly get any leftovers.	• Cost per serving may be higher than homemade.
• Could have a variety of items especially for inexperienced cooks.	• Convenience foods are typically high in calories, fat, saturated fat, sugar, salt, and trans-fats.
• Faster presentation and easy cleaning up.	• They tend to lack freshness in fruits and vegetables.
• Less spoilage and waste occur with packaged convenience foods.	• Improper handling can lead to food spoilage.
• Transportation of packaged foods is cheaper especially in concentrated form.	• Canned foods tend to develop specific flavour.
• Cost efficient for mass production and distribution.	• High initial investment is required
• Ready to eat cereal and instant breakfast difficult to prepare at home because of its expensive product technology used in preparation.	• More expensive than fresh foods.
• Reliable and consistent in quality	• May be less meat, fish, or cheese than included in homemade versions.
• Available in all seasons	
• Ethnic and exotic meals are readily available.	
• Saving space and labor.	
• Have nutritive value equal to or higher than fresh foods.	
• Reduction in preparation time, and the wastage	
• Allows greater variety and quality.	
• Very easy to handle and attractive.	

Importance and value	Need
• Traditional foods are available at commercial level, which are prepared by cottage industries and by multinational companies.	• Increased education and employment opportunities for women.
• Supplementation with protein rich sources and preparation of ready to eat foods would correct nutritional inadequacy and provide variety.	• Large number of employed couples.
• Instant mixes are needed for all segments of population, including armies, railways and even patients.	• Increased industrialization and urbanization.
• The convenience foods meet the urgent needs of situation of offering hospitality to unexpected guests and a busy house wife.	• Large floating populations due to promotion of tourism
• Changing lifestyle has brought about changes in eating habits that has great demand for convenience foods in Indian market.	• Better wages and consequently higher surplus incomes.
• Desire for more leisure time and demand for foreign or sophisticated dishes inspired by the media and increased trend.	• Changing lifestyle and food habits of middle income groups.
• Meets the desire to taste new products.	• Better awareness and availability of convenience foods.
	• Lack of time, busy and fast moving life.

Critical considerations	Scope
• Several groups have cited the environmental harm of single serve packaging due to increased usage of plastics that contribute to solid waste in landfills.	• The target group for convenience foods is considered as 20–60 years due to their buying power.

Contd.

Critical considerations	Scope
• Health organizations have spoken out about high levels of salts, fats and preservatives in those products that contribute to obesity epidemics in western and developing nations.	• In recent years there is an increase in number of working women, changing lifestyle, and increase in nuclear and double income families. • Influence of media, increase in foreign travel, integration of ethnic foods into the local food habits and desire for quality foods. There is an increased need for convenience foods due to all these.

Classification

Convenience foods can be categorised in various ways, but are usually classified depending on the level of preparation necessary to prepare the final product.

I. **Classification based on level of preparation and processing**

1. *Basic*: Dried, frozen or canned foods with one or very few ingredients, e.g. instant potatoes, frozen juice concentrates, canned vegetables, etc.

2. *Complex*: A food mix made up of several ingredients offering reduced process/cooking time, e.g. ready-to-use frosting, frozen waffles, frozen entrees, etc.

3. *Highly processed*: Products that cannot be made at home since they are made using advanced processes and technology, e.g. carbonated beverages, instant breakfasts, ready-to-eat cereals, etc.

II. **Classification based on level of preparation and skill needed**

1. **Ready to eat (RTE) foods:** The foods that can be directly consumed from the package with or without warming, thawing and without preparation are called RTE foods.

   o **Dairy snacks**: Processed cheese, cheese spread, butter and ghee.

   o **Dairy sweets**: *Gulab jamuns, kala jamun, Rasgullas, pedhas and burfis.*

- o **Other sweets**: *Sohan papdi, sohan halwa, jilebas, Mysore paks, besan laddu* and other sweets.
- o Bakery Products: Biscuits, bread and cakes
- o **Fried snacks:** Chips, wafers, fried legumes and other snacks.
- o **Retort processed foods:** *Paneer* curries, Dal fries, *parathas* can be packed well in retort pouch made of polypropylene for six months. The products can be heated along with pouches and eaten as and when needed.
- o **Frozen foods**: Ice cream, *idli, chicken, kabab,* fruits and vegetables.
- o **Extruded snacks**: Cereal and pulse based, soya based extruded snacks.
- o **Traditional sweet meats**: *Modakas, laddus, madeli, karchikai and holige.*
- o **Adjuncts**: Pickles, dry chutneys, fruit chutney
2. **Ready to use (RTU) foods:**
   - o Foods which need some preparation like cooking, frying and reconstitution before consumption are called ready to use.
   - o **Masalas:** Butter chicken mix, garam masala, *chat masala*, meat *masala*, curry *masala, palav* mix, *puliogare* mix, *rasam* powder, *sambar* powder, ginger and garlic paste.
   - o **Fresh cut vegetables:** Carrots, beans, cabbages and others are washed and cut into slices, cubes and shreds and modified atmosphere packed.
3. **Ready to cook (RTC) foods:** Noodles, instant *idli, dosa* and *rava idli,* mixes.
4. **Ready to fry (RTF) foods:** Papads, fingers chips, wafers, fryums and chicken.
5. **Ready to reconstitute foods:** *Khoa* powder, *kulfi* mix, instant ice cream mix and weaning mixes.
6. **Breakfast cereals**: Cornflakes, wheat flakes, jowar and millet based flakes, pops and extruded cereals.
7. **Canned foods:** Fruits vegetables, pulps, rasagulla, jamun, curries, meat, fish and chicken.

8. **Beverages**
   o **Ready to drink (RTD) beverages:** The drinks that can be directly consumed from container like apple juice, mango drink, strawberry shake and milk-based beverages.
      1. Horlicks and malt shakes are available in tetra packs, with a shelf life of 4 months.
      2. Sweet *lassi* and cold coffee are available with a shelf life of 6 months.
      3. Natural fruit juices in tetra packs are available with a shelf life of a year.
   o **Ready to serve (RTS) beverages:** These beverages need some preparation before serving. The beverages have to be diluted or reconstituted before use. These include fruit concentrates in different flavours, Tropicana, spicy tomato rasam, soup, chicken soup, instant soup powders and instant juice powders like rasna.

The choice between various categories available is dependent on individual needs and conveniences. Given the advent of movement from the Nutrition Transition Model Phase 4 to 5, consumers have become very conscious about healthy food choices. This trend and demand has translated into the availability and innovation of healthy and functional food products.

INFORMATION OF PRODUCTS ON LABELS

With a growing awareness regarding nutritional trends and healthy eating, consumers are demanding more information of product labels and expecting more benefits from the packed, processed foods that they purchase. The same is demanded even from convenience foods. Several types of beneficial foods are now being manufactured and sold in easy-to-serve or ready-to-eat packing formats.

1. **Health promoting foods**—low calories, low fat, low cholesterol, low-sugar.
2. **Functional foods**—antioxidants, anti-ageing, anti-diabetic, heart-healthy and weight loss foods.

3. **Fortified and enriched foods**—vitamin enriched, high calcium.

4. **Foods aimed at food allergies**—gluten-free and lactose-free foods.

The most popular foods in this food processing sector are soya, cereal bran, onion, garlic, many fruits and vegetables and a wide variety of foods with potential health benefits. Given their huge importance in the upcoming food processing market "functional foods" are discussed below.

They have gained a pivotal role with the increasing awareness of the role of diet in disease prevention as these provide health benefits.

Functional foods are those foods which provide nutrition, as well as certain possess some health benefits.

Functional food is also known as "a food that has a component incorporated into it to give a specific medical or physiological benefit, other than a purely nutritional benefit".

In other words, these are the food products having a defined and well-established health claim.

Health claim can relate to components of food or foods themselves. Three types of health claim are:

1. **Generic:** Related to a specific nutrient within the food product to a particular disease or condition
   ○ For example "Diets low in saturated fat and cholesterol and rich in fruits, vegetables and grain products that contain some type of fiber, particularly soluble fiber, may reduce the risk of heart disease, a disease associated with many factors". On the basis of this statement, new products containing one or more of the above ingredients could be developed.

2. **Commodity specific:** Related to the commodity or ingredient.
   ○ For example as for oats, diets high in oat bran/ oatmeal and low in saturated fat and cholesterol may reduce the risk of heart disease. This highlights the health benefits of oat bran and meal. However, it does not indicate in the claim that the product on which the claim is placed is protective.

3. **Product specific:** Related to the entire product on which the claim is placed has a protective effect against a disease.
   o For this type of claim, the product itself, rather than the ingredients or nutrients in it, have to be shown to have the beneficial effect/claim.

In all the above three types of claims, it is essential to provide scientific evidence in support of the beneficial effect of the commodity or ingredient or the product in the prevention or treatment of a condition.

Some functional foods are discussed below:

Cereal and Pulse-based Products

1. Oat products

These days there is an influx of oat products in the markets and as a result in homes of people. It is largely replacing other cereal ingredients.

Oats fulfil admirably the description of a functional food, as one that, in addition to providing all normal attributes of a food-basic substance, pleasing taste and texture—also confers a specific health benefit.

The outer layers of oats are similar to those of other cereals in being a good source of insoluble dietary fiber with the attendant capacity to improve colonic function and possibly reduce the risk of colon cancer. Many other functionally distinct components such as waxes, lignin, phytate, vitamins, minerals and phenolics concentrate in these layers. Some of these compounds are powerful antioxidants and may possess potent pharmacological properties.

The Food and Drug Administration (FDA) of the USA has recently allowed a health claim for an association between consumption of diets high in oatmeal, oat bran, or oat flour and reduced risk of coronary heart disease. This represents the first health claim for a specific food under the *Nutrition Labelling and Education Act* (1990). The overall conclusion from the FDA review was that oats could indeed lower serum cholesterol levels, specifically low-density lipoprotein (LDL) cholesterol, without change in the high-density lipoprotein (HDL) fraction, on this basis, a health claim for reduced heart disease risk was allowed.

2. Wheat bran

The useful role of wheat bran in promoting regularity in colonic function and preventing constipation is generally accepted. In addition, growing research has focused its protective effect against colon and breast cancers.

Amount of fiber in the diet has an effect on colonic function, the type of fiber and its digestibility also play a significant role. Both soluble and insoluble fibers have value in promoting regularity in colonic function, as measured by stool weight and transit time, but they promote regularity via different mechanisms. Insoluble fibers, such as those from wheat bran, are resistant to fermentation by colonic bacteria and increase faecal bulk by retaining water.

Among the different sources of dietary fiber as faecal bulking agents, *wheat bran* is probably the most studied and among the most effective.

From a food processing perspective, the range of particle size in commercially available wheat bran offers many functional benefits. While fiber particle size may affect its colonic effects, the range of particle size typically found in commercially available wheat bran (coarse bran > 1400 µm to very finely ground bran <500 gm) is well within that reported to be associated with faecal bulking effects.

3. Rice bran

Rice bran contains primarily insoluble fiber (cellulose) and soluble fiber (hemicellulose). Insoluble fiber adds bulk to the gastrointestinal (GI) tract in humans causing more frequent stools that pass through the system more quickly, requiring less pressure to expel and absorbing more bile acids, thereby preventing their re-entry into circulation.

This lowers the amount of bile absorption/re-absorption of dietary and or endogenous lipid by the lower intestinal tract and promotes the synthesis of more bile acids from available cholesterol. Lowering serum cholesterol levels in the blood, specifically the low-density lipoprotein (LDL) fraction, aids in cardiovascular health and tends to lessen gallstone formation.

Rice bran is potentially a valuable source of natural antioxidants such as tocopherols, tocotrienols and oryzanols. Increased concern over the safety of synthetic antioxidants like butylated hydroxynisole (BHA) and butylated hydroxytoluene (BHT) has increased the interest in finding effective and economical antioxidants. Antioxidants extracted from rice bran potentially could satisfy this demand.

Defatted rice bran contains an increased percentage of fiber ranging from 35 to 48%, and can be used in speciality high—fiber products and baked goods. Rice bran fractions also possess emulsifying and foaming properties for baked products, meringues and whipped toppings. These fractions reportedly provide other benefits, such as leavening and texture virtues.

4. Soya bean products

Soya bean and its flours are used in the preparation of a variety of fermented and non-fermented products in Asian countries. However in India, food use of soya bean or soya flour is limited to a few extruded and texturized products.

Soya proteins are known to reduce cholesterol levels in hypercholesterolaemic individuals. The effect is greatest on those with the highest starting levels of cholesterol. These findings strongly support the inclusion of soya proteins (soya protein isolate and soya protein concentrate) in a wide variety of common food products.

An average of 17–25 gm/day of soya protein has been found to be effective in lowering serum cholesterol. Epidemiological and animal studies have indicated that there could be a correlation between consumption of soya proteins and certain chronic diseases like breast and prostrate cancer. In India, food grade soya meal (defatted soya flour) is available which can be used in formulation of new food for health benefits.

Fruits and Vegetable Based Products

5. Grapes

The components of grapes and grape products believed to play a significant role in preventing or delaying the onset of diseases including cancer and cardiovascular diseases are the *phenolic compounds.*

These compounds are secondary plant metabolites that contribute in an important manner to the flavour and colour characteristics of grapes, grape juices and wines.

The phenolic compounds of grapes include *phenolic acids, anthocyanins, flavonols, flavan-3-ols* and *tannins.* The flavonoids (C6-C3-C6), which include the anthocyanins, flavonols and flavan-3-ols, are powerful antioxidants, and are found in high concentration in grapes and grape products. These compounds exhibit a wide range of biochemical and pharmacological effects including anti-inflammatory and anti-allergic effects.

From the foregoing, it is evident that the grapes and grape products are rich in phenolic compounds, particularly flavonoids, which have demonstrated a wide range of biochemical and pharmacological effects, including anticarcinogenic, antiathero-genic, anti-inflammatory, antimicrobial and antioxidant activities. The available information suggests that the regular consumption of currently available grape products should have a long-term health benefit. However, for increased concentration of grape phenolics, such as resveratrol, ellagic acid and flavonoids, new food products rich in these phytochemicals need to be developed.

The byproducts of wine-making, grape skins, seeds and cluster stems are rich *in catechins, proanthocyanidins and/or natural antioxidants,* which can then be incorporated in the variety of foods such as breakfast cereals, bakery products and confectionaries.

6. Citrus fruits

A large number of constituents in citrus products have been shown to be capable of preventing or alleviating diseases and promoting health. Vitamins C, E and carotenoids, for instance, are thought to play a role in preventing or delaying the onset of major degenerative diseases of aging such as cancer, cardiovascular disease and cataracts by counteracting oxidative processes. Similarly, several "non-nutrient" components of citrus, including limonoids and flavonoids, appear to inhibit carcinogenesis by acting as blocking and/or suppressing agents.

7. Onion and garlic

Onions *(A. cepa)* and garlic *(A. sativam)* have been used in traditional and folk medicine for over 4000 years. Disorders for which both garlic and onions have been used include asthma, arthritis, arteriosclerosis, chickenpox, the common cold, diabetes, malaria, tumors and heart problems. Modern science has shown that alliums and their constituents have several therapeutic effects, including *antiplatelet aggregation* activity, *fibriolytic activity,* anticarcinogenic effects, antimicrobial activity, anti-inflammatory and anti-asthmatic effects.

Onion and garlic based products are currently marketed in a variety of forms. They include, for onions: dehydrated onion pieces, onion powder, onion flavourings, encapsulated flavours, oleoresins and essential oils, onion salt, pickled onions, canned, frozen and packaged onions; for garlic: dehydrated garlic powder, garlic salt, garlic juice, and garlic flavouring, encapsulated flavours, oleoresins and essential oils.

Oil-based Products

8. Fish liver oils

Fish liver oils originate from the liver of lean white fish such as cod, the body of oily fish such as mackerel and the blubber of marine mammals such as seal. These oils consist of saturated, monounsaturated and polyunsaturated fatty acids (PUFA). There are two classes of PUFA, namely *the omega-3* and *omega-6 families,* which are differentiated from one another based on the location of the double bond from the terminal methyl group of the fatty acid molecule. Unlike saturated and monosaturated fatty acids, which can be synthesized by all mammals, including humans, the PUFA cannot be easily synthesized in the body and it must be provided through the diet.

The unique feature that differentiates lipids of marine species from those of land animals is the presence of long-chain PUFA, namely eicosapentaenoic acid (EPA), docosahexaenoic acid (DHA) and to a lesser extent, docosapentaenoic acid (DPA).

Consumption of marine oil results in a decrease in plasma lipids by reduced synthesis of fatty acids and low-density lipoproteins. It has also been suggested that marine oils may retard atherogenesis through their effect on platelet function, platelet-endothelial interactions and inflammatory response.

Seeds-based Products

9. Flaxseeds

Flaxseeds are derived from the plant Flax. Flaxseeds occur in two basic varieties: Brown and yellow or golden (also known as golden linseeds). Most types have similar nutritional characteristics and equal number of short-chain omega-3 fatty acids. The exception is a type of yellow flax called solin (trade name Linola), which has a completely different oil profile and is very low in omega-3 FAs.

A 100 g portion of ground flaxseed supplies about 534 calories, 41 g of fat, 28 g of fiber, and 20 g of protein.

Flaxseed sprouts are edible, with a slightly spicy flavour. Excessive consumption of flaxseeds with inadequate water can cause bowel obstruction. In northern India, flaxseed, called *tisi* or *alsi*, is traditionally roasted, powdered, and eaten with boiled rice, a little water, and a little salt.

Whole flaxseeds are chemically stable, but ground flaxseed can go rancid at room temperature in as little as one week. Refrigeration and storage in sealed containers will keep ground flax from becoming rancid for a longer period. Milled flax is stable to oxidation when stored for nine months at room temperature if packed immediately without exposure to air and light and for 20 months at ambient temperatures under warehouse conditions.

Three natural phenolic glucosides, secoisolariciresinol diglucoside, p-coumaric acid glucoside, and ferulic acid glucoside, can be found in commercial breads containing flaxseed.

In a 100 gram serving, flaxseed contains high levels (> 19% of the daily value, DV) of protein, dietary fiber, several B vitamins, and dietary minerals. Flaxseeds are especially rich in thiamine, magnesium, and phosphorus (DVs above 90%).

As a percentage of total fat, flaxseeds contain 54% omega-3 fatty acids (mostly ALA), 18% omega-9 fatty acids (oleic acid), and 6% omega-6 fatty acids (linoleic acid); the seeds contain 9% saturated fat, including 5% as palmitic acid. Flaxseed oil contains 53% 18:3 omega-3 fatty acids (mostly ALA) and 13% 18:2 omega-6 fatty acids.

Consuming flaxseed or its derivatives reduces total and LDL-cholesterol in the blood, with greater benefits in women and those with high cholesterol.

10. Chia seeds

Chia seed is obtained most commonly from *Salvia hispanica* of the *Lamiaceae* family. Other plants referred to as 'chia' include 'golden chia' (*Salvia columbariae*) and the flowering herbaceous perennial *Salvia polystachya*, which is rarely cultivated and the seeds are not used. The seeds of *Salvia columbariae* are used medicinally and for food. Chia seeds served as a staple food of the Nahuatl (Aztec) cultures of Central Mexico. Ground or whole chia seeds are still used in Paraguay, Bolivia, Argentina, Mexico, and Guatemala for nutritious drinks and food.

Chia seeds is rich in omega-3 fatty acids since the seeds yield 25–30% extractable oil, including α-linolenic acid. The composition of the fat of the oil may be 55% ω-3, 18% ω-6, 6% ω-9, and 10% saturated fat.

Typically, chia seeds are small ovals with a diameter of approximately 1 mm (0.039 in). They are mottle-coloured with brown, gray, black, and white. The seeds are hydrophilic, absorbing up to 12 times their weight in liquid when soaked. While soaking, the seeds develop a mucilaginous coating that gives chia-based beverages a distinctive gel texture.

A 100 g serving of chia seeds is a rich source of the B vitamins thiamine and niacin (54% and 59%, respectively of the daily value (DV)), and a good source of the B vitamins riboflavin and folate (14% and 12%, respectively).

The same amount of chia seeds is also a *rich* source of the dietary minerals calcium, iron, magnesium, manganese, phosphorus, and zinc (more than 20% DV). Chia seeds may be added to other foods as a topping or put into smoothies, breakfast cereals, energy bars, granola bars, yogurt, tortillas, and bread. In 2009, the European Union approved chia seeds as a *novel food*, allowing chia to be 5% of a bread product's total matter.

They also may be made into a gelatin-like substance or consumed raw. The gel from ground seeds may be used to replace as much as 25% of the egg content and oil in cakes while providing other nutrients.

Although preliminary research indicates potential health benefits from consuming chia seeds, this work remains sparse and inconclusive. In a 2015 systematic review, most of the studies did not demonstrate a statistically significant effect of chia seed consumption on cardiovascular risk factors in humans.

ANTIOXIDANTS AND HEALTH PROMOTION

The primary biological role of antioxidant is in preventing the damage that reactive free radical can cause to cells and cellular compounds.

Free radical is a group of atoms that behave like a unit, e.g. carbonate radical, (CO_3^{2-}), nitrate radical (NO_3), and Methyl radical (CH_3^-). *Free radical contains one or more unpaired electrons.*

Human body naturally produces free radicals as it metabolizes oxygen. Reactive free radicals are able to produce metabolic disturbances and to damage membrane structures in a variety of ways. This may lead to cardiovascular disease, cancer and other health problems.

The current dietary recommendation to increase fruit and vegetable consumption is one which is widely perceived as health-promoting. Consistent epidemiological links worldwide between high fruit and high vegetable consumption and a greater life expectancy warrant more emphasis given to this particular dietary recommendation.

Fruits and vegetables are the rich sources of the antioxidants, vitamin C, vitamin E, various carotenoids, flavonoids, isoflavones, organo-sulphur compounds, copper, manganese and magnesium and may also contribute to pools of endogenously produced antioxidants such as *ubiquinol.* Fruits and vegetables, however, are not the only dietary source of antioxidants. Other rich sources of vitamin E include nuts and seeds, wholegrain breakfast cereals, wholemeal bread, eggs, margarine, vegetable oils and dairy products.

ENVIRONMENTAL AND HEALTH CONCERNS

Types of convenience foods can vary by country and geographic region. Some convenience foods have received criticism due to concerns about nutritional content and how their packaging may increase solid waste in landfills. Various methods are used to reduce the unhealthy aspects of commercially produced food and fight childhood obesity.

Several groups have cited the environmental harm of single serve packaging due to the increased usage of plastics that contributes to solid waste in landfills. Due to concerns about obesity and other health problems, some health organizations have criticized the high fat, sugar, salt, food preservatives and food additives that are present in some convenience foods.

In most developed countries, 80% of consumed salt comes from industry-prepared food (5% come from natural salt; 15% comes from salt added during cooking or eating).

Health effects of salt concentrate on sodium and depend in part on how much is consumed. A single serving of many convenience foods contains a significant portion of the recommended daily allowance of sodium. Manufacturers are concerned that if the taste of their product is not optimized with salt, it will not sell as well as competing products. Tests have shown that some popular packaged foods depend on significant amounts of salt for their palatability.

Convenience food, or processed food, is commercially prepared food created as an easy way to get and consume. Most convenience foods provide a little to no nutritional value and have excessive amounts of sodium, sugar, and saturated fats. While everyone should avoid these types of foods, it is highly recommended for individuals with health conditions like heart disease, hypertension, or diabetes to avoid these foods altogether.

Processed foods are also loaded with preservatives (MSG for example), unnatural colouring, added flavouring, and other unappetizing substances. If consumed regularly over time, such foods can quickly begin to harm a person's health, which can contribute to serious health issues, for example, obesity, diabetes, heart disease, cancers, and strokes.

The craving for processed food is more contributed to its added flavours and textures than the food itself. Convenience foods are developed with excessive amounts of salt and fats to give you a sensory overload to get you hooked, thus addicted to the need for its satisfying taste.

It can take a few weeks to months to detox the taste buds, but it is better to do a gradual reduction in salt intake from processed foods. This way the taste buds can relearn the tastes of foods in their natural form at a more effective, slower pace.

The body's ability to digest these foods can be difficult, as convenience food become modified when processed. Processed foods typically lack micronutrients which are required in trace amounts for the normal growth and development of living organisms, like our bodies. Micronutrients, more commonly known as vitamins and minerals, play an important role in health by keeping the internal systems functioning properly. They include such vitamins and minerals as vitamin A, vitamin C, vitamin D, vitamin E, vitamin K, and B-vitamins, and minerals including magnesium, iodine, sodium, zinc, and copper.

Below is a list of popular convenience/processed foods:
o Packaged chips
o Canned vegetables
o Bread
o Commercialized fruit juices
o Salt, sugar, flour
o Frozen meals/pre-packaged foods
o Items from fast-food menus

Below is a list of healthy whole foods to replace some of these popular convenience foods:
o Instead of packaged chips like potato chips, make own chips or fries from whole organic russet potatoes or sweet potatoes at home.
o Buy fresh, seasonal produce instead of canned. Many canned veggies are high in added sodium, even when the label says "low sodium." If still to be used, these must be properly drained and washed with water before use.

○ Use a juicer or squeeze fruit juice at home. Doing this will not only provide the body with rich nutrients, but it will also give a healthy boost of energy to go about the day.

○ Opening a pre-packaged frozen meal at home still makes it a processed meal. Many households these days are crunched for time and it seems as though cooking at home has been shoved to the sidelines. Instead of buying frozen meals, set aside some time to make healthy homemade meals to portion out into freezer containers.

Fast Food

Fast food refers to food that can be prepared and served quickly; it could be small scale or mass produced. Fast food restaurants usually have a walk up counter and/or drive-thru window where one orders and picks up food without having to wait long.

According to the National Institutes of Health (NIH), fast foods are quick alternatives to home-cooked meals. They are also high in saturated fat, sugar, salt and calories.

Although they are convenient for the consumer, the food is often made with cheaper ingredients such as high fat meat, refined grains, and added sugar and fats, instead of nutritious ingredients such as lean meats, whole grains, fresh fruits, and vegetables.

The food is typically less nutritionally valuable compared to other foods and dishes. While any meal with low preparation time can be considered fast food, typically the term refers to food sold in a restaurant or store with preheated or precooked ingredients, and served to the customer in a packaged form for take-out/take-away.

Eating too much fast food has been linked to, among other things, colourectal cancer, obesity and high cholesterol.

TRACING THE ROOTS OF FAST FOOD

In Ancient Rome and India, cities had street stands—a large counter with a receptacle in the middle from which food or drink would have been served. It was during post-World

War II American economic boom that Americans began to spend more and buy more as the economy boomed and a culture of consumerism bloomed. As a result of this new desire to have it all, coupled with the strides made by women while the men were away, both members of the household began to work outside the home. Eating out, which had previously been considered a luxury, became a common occurrence, and then a necessity. Workers, and working families, needed quick service and inexpensive food for both lunch and dinner. This need is what drove the phenomenal success of the early fast food giants, which catered to the family on the go.

Fast food outlets are *take-away* or *take-out* providers that promise quick service. Such fast food outlets often come with a 'drive-through' service that lets customers order and pick up food from their vehicles. Others have indoor or outdoor seating areas where customers can eat on-site. In recent times, the boom in IT services has allowed customers to order food from their homes through their smart phone apps.

Nearly from its inception, fast food has been designed to be eaten "on the go," often does not require traditional cutlery, and is eaten as a finger food. Common menu items at fast food outlets include burger, vegetable chowmein, momo, fish and chips, sandwiches, pitas, hamburgers, fried chicken, french fries, onion rings, chicken nuggets, tacos, pizza, hot dogs, and ice cream.

Street Vendors

Traditional street food is available around the world, usually though small and independent vendors operating from a cart, table, portable grill or motor vehicle. Common examples include

o Indian street fast food
  - Poori-bhaji
  - Samosa
  - Bati-chokha
  - Chole-kulcha
  - Bhelpuri
  - Chowmein

- Burger
- Tikki
- Pani puri
o Vietnamese
  - Noodle vendors
  - Vegetable stew
o Middle Eastern
  - Falafel stands
  - Doner kebabs
  - Pita bread
  - Baklava
o New York City
  - Hot dog carts
  - Taco trucks
  - Salad and fruit juice carts
  - Cupcakes and baked items
o Philippine
  - Turo-Turo vendors

Commonly, street vendors provide a colourful and varying range of options designed to quickly captivate passers-by and attract as much attention as possible.

INDIAN FAST FOOD

The fast food industry in India has evolved with the changing lifestyle of the young Indian population. The sheer variety of gastronomic preferences across the regions, hereditary or acquired, has brought about different modules across the country. It may take some time for the local enterprise to mature to the level of international players in the field.

Many of the traditional dishes have been adapted to suit the emerging fast food outlets. The basic adaptation is to decrease the processing and serving time. For example, the typical meal which called for being served by an ever alert attendant is now offered as a Mini-Meal across the counter. In its traditional version, a plate or a banana leaf was first laid down on the floor or table. Several helpers then waited on the diner, doling out different dishes and refilling as they got over in the plate.

In the fast food version, a plate already arranged with a variety of cooked vegetables and curries along with a fixed quantity of rice and Indian flatbreads is handed out across the counter against a prepaid coupon. The curries and breads vary depending on the region and local preferences. The higher priced ones may add a sweet to the combination. Refills are generally not offered.

The diversity of Indian cuisine poses logistical problems when it comes to handling. Hence it is common to serve different cuisines at different counters within the same premises. Presence of a large vegetarian population, who eschew non-vegetarian food, has given rise to outlets which exclusively serve vegetarian fast food. Also, different variety of food may be served depending on the times of the day. Beverages such as coffee, tea, soft drinks and fruit juices may also be served in such outlets. Some outlets may additionally have specially designed counters for ice-cream, chaats, etc.

Popular formats of fast food business in India have the following features in common:

o Wide opening on the road side
o Easy to maintain and durable décor
o A cash counter where food coupons are sold
o A food delivery counter which invariably is granite topped
o Additional counters for ice creams, chaats, beverages, etc.
o A well-fitted kitchen located so as to be visible to the customers
o Tall tables, usually of stainless steel, where one can eat while standing
o A drinking water fountain adorned with a water filter
o Rust-proof and non-breakable crockery

Most of the fast food outlets in India are stand alone establishment, a few of them having more than one branch.

Darshini

One of Bangalore's restaurateurs, Mr Prabhakar, opened an outlet called Upahara Darshini in mid-1980s. The novelty was that the food is cooked just behind the serving counter, visible to the customers, and one has to eat while standing by placing the food on tall tables. It is a self-service place where one has to buy a coupon before eating. It offered typical south Indian snacks at highly affordable prices resulting in an instant hit with the office goers as well as students. The size and the enclosed design of the eating space and consequent spilling over of the eaters onto the footpath during the busy hours indicates that he did not anticipate the level of success. This issue is addressed by those who copied the module by keeping the entire face of the outlet open to the road. It would not be wrong to say that this was a trend setter and its format, described earlier, is even to this day replicated by other across south India. The popularity of this business module can be gauged from the fact that many restaurants which adopt this format have "Darshini" as suffix in their names.

Food Courts

Another concept of fast food that is becoming popular is that of Food Courts. It is like putting together a number of Darshinis serving different cuisines under one roof. Here also one has to purchase coupons and collect the food from one of the several counters. Each one of these counters serves specific variety of food and may be owned by different individuals or caterers. Food courts are normally located on much bigger premises and may provide seating facility in addition to the stand and eating arrangement. Typically one entrepreneur owns or takes on lease the entire premises and promotes the place under one name. He then lets out individual counters to different independent operators to offer different menu. Internal competition is avoided by not allowing more than one counter to offer similar food.

Several international fast food chains like Kentucky Fried Chicken, McDonald's, and Barista Coffee have their outlets in major cities. Café Coffee Day, again a brainchild of Bangalore-based businessman, is the only Indian chain which boasts of hundreds of outlets and is present across India. But then it is classified more as a coffee shop than a fast food place.

Now local chains coupled with numerous foreign fast foods have sprung up in India, leading to many websites not only catering to the curated list of foods, restaurants and reviews but also giving option to book and get it delivered at your doorsteps.

Variety of options available in the regional cuisines of India.

West India	East India	South India	North India
• Dhokla	• Momos	• Idli	• Poori-sabji
• Samosa	• Spring rolls	• Vada	• Pakodas
• Sev Puri	• Chowmein	• Dosa	• Rajma-chawal
• Khandvi	• Fried rice	• Upma	• Chole-kulcha
• Gathiya	• Egg roll	• Pongal	• Bonda
• Poha	• Noodle soup	• Vangibath	• Parantha
• Jalebi	• Fish chips	• Curd rice	• Chaat
		• Dahi vada	
		• Bonda	

FAST FOOD "GOOD OR BAD": A DISCUSSION

There is no such thing as a "bad" food, but there are some foods one should try not to have on a regular basis. Because fast food is high in sodium, saturated fat, trans fat, and cholesterol, it should not be eaten often. Eating too much over a long period of time can lead to health problems such as high blood pressure, heart disease, and obesity. People also often drink soda when they eat fast food which adds "empty" calorie to the meal.

Fast-food chains have come under criticism over concerns ranging from claimed negative health effects, alleged animal cruelty, cases of worker exploitation, and claims of cultural degradation via shifts in people's eating patterns away from traditional foods.

The intake of fast food is increasing worldwide. A study done in Jeddah has shown that current fast food habits are related to the increase of overweight and obesity among adolescents in Saudi Arabia. In 2014, the World Health Organization published a study which claims that deregulated food markets are largely to blame for the obesity crisis, and suggested tighter regulations to reverse the trend. In America local governments are restricting fast food chains by limiting the number of restaurants found in certain geographical areas.

To combat criticism, fast food restaurants are starting to offer more health-friendly menu items. In addition to health critics, there are suggestions for the fast food industry to become more eco-friendly. The chains have responded by "reducing packaging waste".

Despite so much popularity, fast foods and fast food chains have adverse impacts not only on the job and social skills, but on the health and academic performance of students.

Many fast food chains have some healthy options to choose from. For example, some chains no longer serve foods with trans fat, and many have menu items that contain fruits and vegetables.

These foods are often high in calories yet offer a little or no nutritional value. When fast food frequently replaces nutritious foods in diet, it can lead to poor nutrition, poor health, and weight gain. Tests in lab animals have even shown a negative effect in short duration diets. Being overweight is a risk factor for a variety of chronic health problems including heart disease, diabetes, and stroke.

A 2013 study published in JAMA Pediatrics showed that children and adolescents take in more calories in fast food and other restaurants than at home. Eating at a restaurant added between 160 and 310 calories a day. The effects of fast food on the human body are discussed below:

1. **Digestive and cardiovascular systems**
    a. Many fast foods and drinks are loaded with carbohydrates and, consequently, a lot of calories. The digestive system breaks down carbohydrates into sugar (glucose), which it then releases into bloodstream. The pancreas respond by releasing insulin, which is needed to transport sugar to cells throughout the body. As the sugar is absorbed, blood sugar levels drop prompting pancreas to release another hormone called glucagon. Glucagon tells the liver to start making use of stored sugars.
    b. When everything is working properly, blood sugar levels stay within a normal range. When high

amounts of carbohydrates are ingested, it causes a spike in the blood sugar. That can alter the normal insulin response. Frequent spikes in blood sugar may be a contributing factor in insulin resistance and type 2 diabetes.

2. **Sugar and fat**
   a. Added sugars have no nutritional value but are high in calories. According to the American Heart Association, most Americans take in twice as much sugar as is recommended for optimal health. All those extra calories add up to extra weight, which is a contributing factor for getting heart disease. With shifting lifestyle patterns in developing countries as well, diet patterns are tilting more towards this trend.
   b. Trans fats are a manufactured fat with no extra nutritional value. They are considered so unhealthy that some countries have banned their use. Often found in fast food, trans fats are known to raise LDL cholesterol levels. They can also lower HDL cholesterol, which is the so-called good cholesterol. Trans fats may also increase the risk of developing type 2 diabetes.

3. **Sodium**
   a. Too much sodium causes the body to retain water, causing to feel bloated. Sodium also can contribute to existing high blood pressure or enlarged heart muscle. In congestive heart failure, cirrhosis, or kidney disease, too much salt can contribute to a dangerous build-up of fluid. Excess sodium may also increase the risk for kidney stones, kidney disease, and stomach cancer.
   b. High cholesterol and high blood pressure are among the top risk factors for heart disease and stroke.

4. **Respiratory system**
   a. Obesity is associated with an increase in respiratory problems. Even without diagnosed medical conditions, obesity may cause episodes of shortness of breath or

wheezing with little exertion. Obesity also can play a role in the development of sleep apnea, a condition in which sleep is continually disrupted by shallow breathing and asthma.

b. A recent study published in the journal Thorax suggests that children who eat fast food at least three times a week are at increased risk of asthma and rhinitis, which involves having a congested, drippy nose.

5. **Central nervous system**

a. A study published in the journal Public Health Nutrition showed that eating commercial baked goods (biscuits, pastries, doughnuts, croissants, bran muffins) and fast food (pizza, hamburgers, and hot dogs) may be linked to depression. The study determined that people who eat fast food are 51 percent more likely to develop depression than those who eat a little to no fast food. It was also found that the more fast food study participants consumed, the more likely they were to develop depression.

b. A junk food diet could also affect the brain synapses and the molecules related to memory and learning, according to a study published in the journal Nature.

6. **Skin and bones**

a. Chocolate and greasy foods are often blamed for acne, but they are not the real culprits. According to the Mayo Clinic, because foods that are high in carbohydrates increase blood sugar levels, they may also trigger acne.

b. The study in Thorax showed a higher risk of eczema (inflamed, irritated patches of skin) among children with a diet high in fast food.

c. When one consumes foods high in carbohydrates and sugar, bacteria residing in the mouth produce acids. These acids can destroy tooth enamel, a contributing factor in dental cavities.

d. Excess sodium may also increase the risk of developing osteoporosis.

In order to strike a balance with consumption of fast foods, below are some guidelines:

○ **Go light on the toppings**
- Choose oil-based dressings such as Italian or balsamic vinaigrette or lemon vinaigrette instead of creamy salad dressings, which are high in saturated fat.
- Use mustard or ketchup instead of mayonnaise or "special sauce".
- When ordering pizza, add veggies instead of meat, and get thin crust instead of deep dish.
- Top your sandwiches with veggies such as onions, lettuce, and tomatoes instead of patty, meat or extra cheese.
- Do not add more salt to your meal. Salt is a major contributor to high blood pressure and heart disease and fast food tends to be loaded with it.

○ **Know how the food is made:**
- Choose foods that are broiled, steamed, or grilled instead of fried.
- Choose soups that are not cream based.
- Dishes labelled deep-fried, pan-fried, basted, breaded, creamy, crispy, scalloped, Alfredo, or in cream sauce are usually high in calories, unhealthy fats, and sodium.
- When ordering a sub or sandwich, select lean meats such as turkey or grilled chicken instead of items such as burgers, steak, or cheese sandwiches.
- Ask for sauces or dressings that come with meals to be served on the side and use just a small amount.

○ **Practice portion control:** Meal portions today are almost twice the size that they were 30 years ago.
- Order smaller entree portions.
- When ordering a side dish such as french fries, order a small, or kid sized portion. Never super-size anything; these options pack in an even larger amount of calories and fat.
- Many fast food restaurants advertise value deals for larger portions of food. These foods may come in what is called a "value box", a combo pack, or just be

a larger portion for a cheap price. It is better to split or eat a little at a time to save on piling up excess calories.

- Even if a fast food restaurant uses healthy ingredients, they still usually give a lot more food than needed. Eating too much of any kind of food can lead to weight gain. Watch portions, even when eating healthier fast foods such as salads, sandwiches, and soup.

o **Make the swap:** Look for healthier side options for meals.

- Have a salad, fruit or soup instead of fries.
- Choose water, low-fat milk, or diet sodas instead of regular sodas, fruit drinks, or milkshakes, which can be a huge source of calories and sugar.
- Instead of a slice of pie or cookie for dessert try fruit and yogurt.

Nutrition Labelling

The FDA (Federal Food and Drug Administration) has proposed labelling requirements for all fast food restaurants. From December 2016, restaurants with twenty or more locations will be required to post the calorie content of foods on the menu. Some restaurants have already started to do this.

Remember that just because something might be lower in calories than another item, that does not necessarily mean that it is "healthier." For example, french fries might be lower in calories than a grilled chicken sandwich, but you would be better off picking the sandwich because it is lower in fat and has more protein.

Most fast food and restaurant chains also offer nutrition information online. Many of these menus are now interactive as well, so one can preview the plate and modify it to be more nutritious.

"The Ministry of Health and Family Welfare had, on September 19, 2008, notified the Prevention of Food Adulteration (5th Amendment) Rules, 2008, mandating packaged food manufacturers to declare on their product labels nutritional information and a mark from the FPO or Agmark (companies that are responsible for checking food products) to enable consumers make informed choices while purchasing. Prior to this amendment, disclosure of nutritional information was largely voluntary though many large manufacturers tend to adopt the international practice."

It is high time that these interventions are legislated and executed for fast food chains in India as well. Policy decisions should be enforced effectively in India and restaurants—fast food chains penalised for their failure to comply.

Chapter 9

Quick Cooking Products

Given the shifts in lifestyle and dietary habits, as discussed in detail in earlier chapters, food companies have come up with quick cooking alternatives to conventional food products. Seizing the opportunity of the rising demands of such products and advancements in food product development research, food companies are flooding the market with quick cooking food products.

Putting it simply, quick cooking products are those food products which require lesser cooking time and effort than their conventional counterparts. They are pre-processed and partially cooked products, thus reducing the cooking time when in use by the consumer. They come with accompaniments (mostly in dehydrated form, to be re-hydrated while cooking) such as dehydrated vegetable chunks, croutons, sauces, pre-mixed spice/flavour sachets, dried cream sauce powders, etc. One just needs to cook them with a measured quantity of water either on stove top or microwave or conventional oven. They are usually beautifully packed in boxes, retort pouches or plastic casing.

Some examples of quick cooking food products are discussed below.

1. Instant Oats or Quick cooking Oats

Quick cooking oats are rolled oats that have been coarsely chopped into smaller pieces to allow them to cook more quickly than regular oatmeal. Usually, they are about 1/4 or 1/3 the size of a regular rolled oatmeal flake. They are larger than instant oatmeal, which tends to have a very powdery consistency.

Instant oats are pre-cooked, dried, and then rolled and pressed slightly thinner than rolled oats. They cook more quickly than steel-cut or rolled oats, but retain less of their texture, and often cook up mushy. Also referred to as quick oats, instant oats are the most processed of the three oat varieties, viz. rolled oats, steel-cut oats and instant oats. These are discussed below.

All oats start off as oat groats—the whole, unbroken grains. Before being processed into any other variety of oats, groats are usually roasted at a very low temperature. This not only gives the oats their nice toasty flavour, but the heat also inactivates the enzyme that causes oats to go rancid, making them more shelf-stable.

The difference between steel-cut, rolled, and instant oats is simply how much the oat groat has been processed which affect their texture and cooking time.

i. **Steel-cut oats:** Also referred to as Irish or Scottish oats, this variety is made when the whole groat is cut into several pieces, rather than rolled. Steel-cut oats look almost like rice grain that has been cut into pieces. This variety takes the longest to cook, and has a toothsome, chewy texture that retains much of its shape even after cooking. Due to its texture, rolled or instant oats do not make a good substitute for steel-cut oats.

ii. **Rolled oats or old-fashioned or whole oats:** Rolled oats are traditionally oat groats that have been dehusked and steamed, before being rolled into flat flakes under heavy rollers and stabilized by being lightly toasted.

Rolled oats look like flat, irregularly round, slightly textured discs. When processed, the whole grains of oats are first steamed to make them soft and pliable, then pressed to flatten them.

Rolled oats cook faster than steel-cut oats, absorb more liquid, and hold their shape relatively well during cooking. In addition to be heated for a warm breakfast bowl, rolled oats are commonly used in granola bars, cookies, muffins, and other baked goods.

 iii. **Instant oats** *can be used in place of rolled oats, although the cook time will be much less, and the final dish will not have as much texture.*

  Rolled oats can be used in place of instant oats, although it will require more cook time, and the final dish will have more texture.

  Thick-rolled oats are large whole flakes, and *thin-rolled oats* are smaller, fragmented flakes. Oat flakes that are simply rolled whole oats without further processing can be cooked and eaten as *"old-fashioned" porridge oats*, but more highly fragmented and processed rolled oats absorb water much more easily and therefore cook faster, so they are sometimes called *"quick" or "instant" porridge oats*. Rolled oats are most often the main ingredient in granola and muesli.

  Rolled oats can be further processed into coarse powder, which, when cooked, becomes a thick broth. Finer oatmeal powder is often used as baby food.

These above constituents are used principally to make oatmeal.

 o Oatmeal is made of hulled oat grains—groats—that have either been ground, steel-cut, or rolled.

 o Ground oats are also called "white oats".

 o Steel-cut oats are known as "coarse oatmeal" or "Irish oatmeal" or "pinhead oats".

 o Rolled oats can be either thick or thin, and may be "old-fashioned" or "quick" or "instant".

 o The term "oatmeal" is often used in the US, Australia, and parts of Canada while it is known as porridge in other parts globally.

Processing

1. The oat grains are dehusked by impact, then heated and cooled to stabilize the oat groats—the seed inside the husk. The process of heating produces a nutty flavour in the oats.

 a. These oat groats may be milled to produce fine, medium or coarse oatmeal.

b. Steel-cut oats may be small and contain broken groats from the dehusking process (these bits may be steamed and flattened to produce *smaller* rolled oats).

c. Rolled oats are steamed and flattened whole oat groats.

   i. *Old-fashioned oats* can be thick and take a while to boil to make porridge.

   ii. Quick-cooking rolled oats (*quick oats*) are cut into small pieces before being steamed and rolled. *Instant oatmeal* is precooked and dried, often with a sweetener, such as sugar, and flavourings added.

Nutritional Value

Steel-cut, rolled, and instant oats all have similar nutritional profile since they are all made from whole oat groats but depending on kind of processing.

o Claimed to lowers blood cholesterol, because of its beta-glucan content

o Claimed to reduce the risk of heart disease, control weight, lower high blood pressure, and even shrink the threat of certain cancers

o Rolled oats have long been a staple of many athletes' diets, especially weight trainers, because of their high content of complex carbohydrates and water-soluble fiber that encourages slow digestion and stabilizes blood glucose levels.

o Oatmeal porridge also contains more B vitamins and calories than other kinds of porridges.

o Full of protein, iron and complex carbohydrates

o Whole oats are an excellent source of thiamine, iron, and dietary fiber.

o Whole oats are also the only source of antioxidant compounds known as **avenanthramides**; these are believed to have properties which help to protect the circulatory system from arteriosclerosis.

Having said that unless oats are picked straight off the stalk, oatmeal is a processed food, so depending on the amount of processing, the nutritional value can be compromised.

Certain popular oatmeal brands and varieties are discussed below.

Quaker weight control—maple and brown sugar instant oatmeal

(*Single serving*: Calories 160, total fat 3 g, sodium 310 mg, dietary fiber 6 g, sugar 1 g)

Ingredients: Whole grain rolled oats, whey protein isolate, maltodextrin, natural and artificial flavours, salt, oat flour, calcium carbonate, guar gum, caramel colour, soy lecithen, acesulfame potassium, sucralose, niacinamide, vitamin A palmitate, reduced iron, pyridoxine hydrochloride, riboflavin, thiamin mononitrate, folic acid.

Reading the nutrition label
- There are three artificial sweeteners in the ingredients list: Maltodextrin, acesulfame potassium, and sucralose—this explains the one gram of sugar on the NFL.
- Maltodextrin is the third ingredient listed, making it one of the main ingredients.
- Other synthetic ingredients include artificial flavours, caramel colouring and calcium carbonate.

Quaker simple harvest maple brown sugar with pecans instant oatmeal

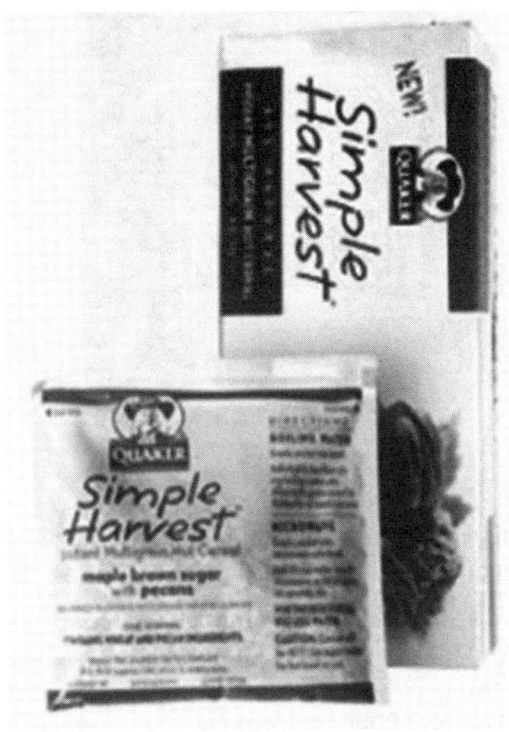

(*Single serving*: Calories 160, total fat 3.5 g, sodium 75 mg, dietary fiber 4 g, sugar 9 g)

Ingredients: Whole grain rolled oats, whole grain rolled wheat, rolled barley, whole grain rolled rye, sugar, pecans, whole flaxseed, oat flour, natural flavours, salt, molasses. Contains wheat and pecan

- o The first four ingredients are easily recognizable: A whole grain blend of oats, wheat, barley and rye.
- o It also includes powerhouse ingredients like flaxseed, oat flour and pecans.
- o There is also a healthy dose of fiber (4 grams) and protein (4 grams), and it is low in sodium.

Mccann's steel cut irish oatmeal

(*Single serving*: Calories 150, total fat 2 g, sodium 0 mg, dietary fiber 4 g, sugar 0 g)

Ingredients: 100% wholegrain irish oats

Notice there is only one ingredient—100 percent whole grains! Nothing is added and nothing is artificial. And, it contains all natural fiber and protein.

2. Instant Breakfast

Instant breakfast typically refers to breakfast food products that are manufactured in a powdered form, which are generally prepared with the addition of milk and then consumed as a beverage.

Some instant breakfasts are produced and marketed in liquid form, being pre-mixed. The target market for instant breakfast products includes consumers who tend to be busy, such as students and working adults.

Carnation-brand instant breakfast was introduced in 1964 in USA. It is a powdered instant beverage that is manufactured with protein, vitamins and minerals and sugar.

3. Instant Rice

Instant rice, also known as minute rice, is rice that has been precooked and dehydrated so that it cooks more rapidly. Regular rice requires 10–15 minutes to cook while instant rice needs anywhere between 5 and 10 minutes. Because it has already been cooked, all that is necessary to prepare instant rice is to simply re-hydrate it with hot water.

Instant rice is made using several methods. The most common method is similar to the home cooking process.

The rice is blanched in hot water, steamed, and rinsed. It is then placed in large ovens for dehydration until the moisture content reaches approximately twelve percent or less.

The basic principle involves increasing moisture of the milled white rice by using steam or water to form cracks or holes in the kernels. The fast cooking properties come from the fact that, when re-cooked, water can penetrate into the cracked grain much more quickly.

Advantages

The major advantage of instant rice is the rapid cooking time—some brands can be ready in as little as three minutes.

Currently, several companies, Asian as well as American, have developed brands which only require 90 seconds to cook, much like a cup of instant noodles.

Disadvantages

o Instant rice is more expensive than regular rice.
o The "cracking" process can lead to a significant increase in broken grains in a package.
o Rice naturally has minerals like phosphorus, magnesium, and potassium. Instant rice has fewer of the calories, carbohydrates, and protein than regular rice. Companies make up for the loss of nutrients by adding their own nutrients such as the B-vitamins, as well as iron.
o Due to its processing, it also loses some of the flavour, but companies compensate by adding herbs and exotic spices and aromas to make it more appetizing.

o The quicker cooking method can result in the rice being less firm in texture than regular rice.

Major brands that produce this kind of product are Minute Rice, Kraft, Rice-A-Roni, Uncle Bens.

4. Instant Noodles

Instant noodles are a precooked and usually dried noodle block, sold with flavouring powder and/or seasoning oil. The flavouring is usually in a separate packet, although in the case of cup noodles the flavouring is often loose in the cup. Some instant noodle products are seal packed; these can be reheated or eaten straight from the packet. Dried noodle blocks are cooked or soaked in boiling water before eating.

The main ingredients used in dried noodles are wheat flour, palm oil, and salt. Common ingredients in the flavouring powder are salt, monosodium glutamate, seasoning, and sugar.

The dried noodle block was originally created by flash frying cooked noodles, and this is still the main method used in Asian countries, but air-dried noodle blocks are favoured in western countries.

Instant noodles were invented by Momofuku Ando (born Go Pek-Hok) of Nissin Foods in Japan. They were launched in 1958 under the brand name Chikin Ramen. In 1971, Nissin introduced Cup Noodles, a dried noodle block in a polystyrene cup (it is referred to as Cup Ramen in Japan). Instant noodles are marketed worldwide under many brand names.

Ando developed the production method of flash frying noodles after they had been made, creating the "instant" noodle. This dried the noodles and gave them a longer shelf life, even exceeding that of frozen noodles.

According to a Japanese poll in the year 2000, "the Japanese believe that their best invention of the twentieth century was instant noodles."

Composition

There are three key ingredients in wheat-based noodles, which are wheat flour, water and salt. Other than the three main ingredients, USDA regulated that noodles from instant ramen noodle soup could also contain palm oil, seasoning, sodium phosphates, potato starches, gums and other ingredients.

Knowing the composition of instant noodles is crucial to understand the physical chemical properties of the product; therefore, the function of each ingredient listed above is specified below.

Flour: Generally, noodles could be made from different kinds of flours, such as wheat flour, rice flour, and buckwheat flour, depending on the various types of product manufacturers want to make.

For instant noodles, flours which have 8.5–12.5% protein are optimal because noodles must be able to withstand drying process without breaking apart, which requires higher amount of protein in flour, and during frying, high protein content could help decrease the fat uptake.

Water: Water is the second most important raw material for making noodles after flour. The hydration of dough will determine the development of gluten structure, which will affect the viscoelastic properties of dough.

The water absorption level for making noodles is about 30–38% of flour weight; if the water absorption level is too high, hydration of flour could not complete, and if the water absorption level is too low, flour dough will be too sticky to handle during processing.

For instant noodles, dehydration is an important step after noodles are made because water can offer growing circumstances for micro-organisms. Depending on dehydration methods, USDA has regulation on moisture content of instant noodles: For instant noodles dehydrated by frying, moisture content could not exceed 8%, and for those dehydrated by methods other than frying, moisture content could not exceed 14.5%.

Salt: Salt is added when making the flour dough to strengthen gluten structures and enhance the sheeting properties of dough, and it can make the noodles softer and more elastic. Salt also offers the basic salty flavour of noodles and can cover some of the off-flavour generated by flour and processing. Another function of salt is to slow the activities of enzymes, such as proteolytic enzymes, which could interrupt the gluten structures, and microbial growth.

Oil: Frying is a common dehydration process for producing instant noodles. Therefore, oil becomes an important component of instant noodles. According to USDA regulation, oil-fried instant noodles should not have fat content higher than 20% of total weight, which means theoretically, amount of oil uptake during frying process could go even higher. Therefore, the high fat content of instant noodles is always the reason why consumers who is pursuing healthy diets would not buy this product.

Palm oil is always chosen as the frying oil for instant noodles due to its heat stability and low cost. However, overall, due to the high fat content and low moisture content, instant noodles are highly susceptible to lipid oxidation, and relatively high amount of preservatives are added. Hence, to avoid the off-flavours and health risking compounds, some instant noodles were dehydrated by ways other than frying to reduce the fat content. According to USDA, un-fried instant noodles should have fat content lower than 3%.

Starches: Potato starches are commonly added in instant noodles to enhance gelling properties and water-holding capacities of noodles. Gelling properties could enhance the elasticity and chewiness of instant noodles, and water holding capacities could improve the smooth and shiny look of noodles after cooking and shorten the cooking time.

Polyphosphates: Polyphosphate is used in instant noodles as additive to improve starch gelatinization during cooking (rehydration) to allow more water retention in the noodles. Normally 0.1% of flour weight phosphate compounds are added to water before mixing and making the dough.

Hydrocolloids: Hydrocolloids such as guar gum are widely used in instant noodles production to enhance water binding capacity during rehydration and shorten the cooking time. Gums are dispersed in water before mixing and making noodles dough.

Nutritional value: A single serving of instant noodles is high in carbohydrates and fat, but low in protein, fiber, vitamins and minerals.

Lead contamination in Nestlé's Maggi brand instant noodles made headlines in India, some 7 times the allowed limit, with several Indian states banning the product as well as Nepal. On June 5, 2015, Food Safety and Standards Authority of India (FSSAI) orders banned all nine approved variants of Maggi instant noodles from India, terming them "unsafe and hazardous for human consumption". It has now been re-instated and available in markets.

In November 2015, Maggi instant noodles made a return in the Indian market after the ban was lifted.

Other popular brands include Top Ramen Smoodles and Cup Noodles manufactured by Indo-Nissin Ltd, Ching's Instant Noodles, Joymee Masala Noodles, AA Nutritions's Yummy, and Wai-Wai, owned by the Chaudhary Group from Nepal. Smith & Jones, Ching's secrets, Tai Pai Noodles & NE Time noodles (Maruti quality Foods Products Pvt. Ltd) are also popular new brands of instant noodles in India. Wai Wai is gaining momentum in North Eastern states, Sikkim and West Bengal.

Local flavours such as masala and chicken tikka dominate. The most popular flavour of Top Ramen is known as "Curry Smoodles"; its flavourings mimic a basic curry, including onion, garlic, coriander, and a curry masala. A package sells for 18 rupees in India. In India, there is also great demand for unflavoured instant noodles; brands such as Bambino and Ching's dominate the market.

Ching's and Smith and Jones are brand of Capital Foods Ltd., (Masala, Curry, Chicken Masala under Smith and Jones brand and Ching's Flavour are Manchurian, Schezwan, Hot Garlic and Chicken Roast Garlic).

ITC launched its Yippee noodle under Sunfeast Brand.

Because of increasing health consciousness, Nestle introduced an instant noodle based on whole wheat grain flour, called Atta Noodles. Instant rice noodles are also available in various flavours. However, Nestle's original "Maggi" masala flavoured noodles continue to be the most successful brand of instant noodles not only in India but also in the United States for Indian Americans. Nestle also has a '2-minute cup noodles' type of maggi known as 'cuppamania', which involves pouring hot water into the prepackaged cup and leaving it to soak and cook.

Foodles, a new instant noodle brand, was launched in late 2010, focusing on health issues with the tagline, ' Noodles without the No '. This range has significantly higher nutrition values compared to other popular brands. It comes in both Multi-grain and Wheat-only forms. The brand is owned by Horlicks.

5. Instant Soup

Instant soup is a type of soup designed for fast and simple preparation. Some are homemade, and some are mass-produced on an industrial scale and treated in various ways to preserve them. A wide variety of types, styles and flavours of instant soups exist. Commercial instant soups are usually dried or dehydrated, canned, or treated by freezing.

Commercial instant soups are manufactured in several types. Some consist of a packet of dry soup stock. These do not contain water, and are prepared by adding water and then heating the product for a short time, or by adding hot water directly to the dry soup mix. Instant soup can also be produced in a dry powder form, such as Unilever's Cup-a-Soup.

Canned (tinned) instant soups contain liquid soup that is prepared by heating their contents. Some canned soups are condensed, and require additional water to bring them to their intended strength, while others are canned in a ready-to-eat, single-strength form. Dr John T Dorrance, an employee with the Campbell Soup Company, invented condensed soup in 1897. In the United States, consumers sometimes utilize condensed soups (without diluting them), as a sauce base. Some instant liquid soups are manufactured in microwaveable containers. Additionally, some instant soups, such as Knorr's Erbswurst, are prepared in a concentrated paste form.

Instant noodle soups such as Cup Noodles contain dried instant ramen noodles, dehydrated vegetable and meat products, and seasonings, and are prepared by adding hot water. Packaged instant ramen noodle soup is typically formed as a cake, and often includes a seasoning packet that is added to the noodles and water during preparation. Some also include separate packets of oil and garnishes used to season the product. Momofuku Ando, the founder of Nissin Foods,[5] developed packaged ramen noodle soup in 1958.

A multitude of instant soup varieties exist. For example, there are several Lipton and Knorr-brand dry instant soups, such as onion, vegetable, tomato beef and cream of spinach. Instant miso soup is generally manufactured in two forms, one as miso paste with preserved vegetable condiments, generally of the shiro (white) kind, and the other as granulated miso. One of the primary uses of dehydrated miso is for the production of instant miso soup. Chicken, beef and seafood/shrimp are the most popular flavours by consumers of ramen noodle instant soups.

Commercially-prepared instant soups are usually dried or dehydrated, canned, or treated by freezing. Some dry instant soups are prepared with thickening ingredients, such as pregelatinized starch, that function at a lower temperature compared to others. Additional ingredients used in commercial instant soups to contribute to their consistency include maltodextrins, emulsified fat powders, sugars, potato starch, xanthan gum and guar gum. Sometimes ingredients used in dry instant soups are ground into fragments, which enables them to be dissolved when water is added. These particulates are sometimes prepared using freeze drying and puff drying.

Freeze drying is a recent dehydration breakthrough-method that is restricted to high-value foods due to the high cost associated with the process. Despite its cost, freeze-drying is very effective in retaining the overall quality and nutritional value of the food through a process called sublimation, where water is removed (evaporated) from food in its frozen state without the transition through the liquid state. Removing water from food prevents spoilage as it does not give an environment that favours the growth of spoilage—causing microorganisms such as bacteria, mould, and yeast. Freeze-drying also reduces the total weight of the food, especially fruit and vegetables that are mostly water, thus making the transportation of these products more efficient. Vegetables used in instant soup mixes often undergo freeze-drying that helps them retain their nutritional value, texture, and flavour. Furthermore, as the food remains rigid during dehydration, subliming water creates holes where the evaporated ice crystals used to be. This allows freeze-dried foods such as vegetables to retain their shape without shrinkage, and these foods rehydrate completely when water is added to the mix that fills the voids left by the subliming water. Due to the reduced water content of the freeze-dried foods that inhibit the growth of microorganisms and prevents enzymatic chemical reactions, these foods are considered shelf-stable and can be kept safe from spoilage for years by preventing the reabsorption of moisture. Freeze-dried foods can be stored in room temperature without the need for refrigeration.

6. Instant Mashed Potatoes

Instant mashed potatoes are potatoes that have been through an industrial process of cooking, mashing and dehydrating to yield a packaged convenience food that can be reconstituted by adding hot water or milk, producing a close approximation of mashed potatoes. They are available in different flavours.

Mashed potatoes can be reconstituted from potato flour, but the process is made more difficult by lumping; a key characteristic of instant mashed potatoes is that it is in the form of flakes or granules, eliminating the chunkiness.

Analogous to instant mashed potatoes are instant poi made from taro and instant fufu made from yams or yam substitutes including cereals. Flattened rice, an instant rice mush, is also much in the same spirit, as more broadly are other instant porridges, formed from flakes, granules, or pearls to avoid lumping. Brands include Smash and Idahoan Foods.

Flaked instant mashed potatoes are most commonly found in stores in the United States and Canada. Granulated forms are generally reserved more for institutional or restaurant use.

The practice of drying and grinding starchy root vegetables for preservation and portability is widely attested around the world, and likely dates back to before the advent of agriculture. Potatoes in particular have been freeze-dried since at least the time of the Inca empire, in the form of chuño; another example is in Japanese Ainu cuisine.

Nutritional value

Instant mashed potatoes have substantially more sodium than fresh potatoes, and much less dietary fiber. In other respects they are similar to mashed fresh potatoes in their nutritional qualities, about two-thirds starch by dry weight, with smaller amounts of protein, dietary fiber and vitamins. The largest difference is the loss of vitamin C, although some products may be enriched to compensate.

Besides being eaten as such there are other uses of instant mashed potatoes like

o Thickening agent in gravy, soups and sauces
o Sourdough starter
o Substitute for bread crumbs
o Sandwich filling
o Parantha filling

Chapter 10

Food Additives and Preservatives

For centuries, ingredients have served useful functions in a variety of foods. Our ancestors used salt to preserve meats and fish, added herbs and spices to improve the flavour of foods, preserved fruit with sugar, and pickled cucumbers in a vinegar solution. Today, consumers demand and enjoy a food supply that is flavourful, nutritious, safe, convenient, colourful and affordable.

Food additives and advances in technology help make that possible. There is a wide variety of food additives available, which are chosen depending on the intended purpose and need of the food product.

Food additives are substances added to food to preserve flavour or enhance its taste and appearance. They can be defined as "any substance the intended use of which results or may reasonably be expected to result—directly or indirectly—in its becoming a component or otherwise affecting the characteristics of any food."

Direct food additives are those that are added to a food for a specific purpose in that food. For example, xanthan gum —used in salad dressings, chocolate milk, bakery fillings, puddings and other foods to add texture—is a direct additive. Most direct additives are identified on the ingredient label of foods.

Indirect food additives are those that become part of the food in trace amounts due to its packaging, storage or other handling. For instance, minute amounts of packaging substances may find their way into foods during storage. Also, traces of nuts can find their way in cereal products, as they might be ground on the same machine at a processing plant.

Natural additives are derived from natural sources (e.g. soybeans and corn provide lecithin to maintain product consistency; beets provide beet powder used as food colouring).

Synthetic additives are ingredients which are not found in nature and therefore must be synthetically produced as artificial ingredients. Also, some ingredients found in nature can be manufactured artificially and produced more economically, with greater purity and more consistent quality, than their natural counterparts. For example, vitamin C or ascorbic acid may be derived from an orange or produced in a laboratory.

Food ingredients are subject to the same strict safety standards regardless of whether they are naturally or artificially derived.

Additives perform a variety of useful functions in foods, some of which are discussed below:

1. Maintain or improve safety and freshness
    i. Preservatives slow product spoilage caused by mold, air, bacteria, fungi or yeast.
    ii. They maintain the quality of the food for a longer duration.
    iii. They help control contamination that can cause foodborne illness, including life-threatening botulism.
    iv. Antioxidants prevent fats and oils and the foods containing them from becoming rancid or developing an off-flavour.
    v. They also prevent cut fresh fruits such as apples from turning brown when exposed to air.
2. Improve or maintain nutritional value: Fiber, vitamins and minerals are added to many foods to make up for those lacking in the diet or lost in processing, or to enhance the nutritional quality of a food.
3. Improve taste, texture and appearance
    i. Spices, natural and artificial flavours, and sweeteners are added to enhance the taste of food.

  ii. Food colours maintain or improve appearance.

  iii. Emulsifiers, stabilizers and thickeners give foods the texture and consistency consumers expect.

  iv. Leavening agents allow baked goods to rise during baking.

  v. Some additives help control the acidity and alkalinity of foods, while other ingredients help maintain the taste and appeal of foods with reduced fat content.

In the US, food and colour additives are more strictly studied, regulated and monitored than at any other time in history. FDA has the primary legal responsibility for determining their safe use. Under the Food Additives Amendment, two groups of ingredients were exempted from the regulation process of FDA. All others have to undergo approval procedures by FDA.

Group I: Prior-sanctioned substances are substances that FDA or USDA had determined safe for use in food prior to the 1958 amendment. Examples are sodium nitrite and potassium nitrite used to preserve luncheon meats.

Group II: GRAS (generally recognized as safe) ingredients are those that are generally recognized by experts as safe, based on their extensive history of use in food before 1958 or based on published scientific evidence. Among the several hundred GRAS substances are salt, sugar, spices, vitamins and monosodium glutamate (MSG).

When evaluating the safety of a substance and whether it should be approved, FDA considers

  o the composition and properties of the substance,

  o the amount that would typically be consumed, immediate and long-term health effects,

  o various safety factors.

E NUMBER

A very important aspect of food additives is the concept of E number. Before studying them in detail one must know the significance and meaning of E number with respect to additives and preservatives.

Table 10.1: Various types of food additives/preservatives/ingredients

Types of ingredients	What they do	Examples of uses	Names found on product labels
Preservatives	Prevent food spoilage from bacteria, molds, fungi, or yeast (antimicrobials); slow or prevent changes in colour, flavour, or texture and delay rancidity (antioxidants); maintain freshness	Fruit sauces and jellies, beverages, baked goods, cured meats, oils and margarines, cereals, dressings, snack foods, fruits and vegetables	Ascorbic acid, citric acid, sodium benzoate, calcium propionate, sodium erythorbate, sodium nitrite, calcium sorbate, potassium sorbate, BHA, BHT, EDTA, tocopherols (vitamin E)
Sweeteners	Add sweetness with or without the extra calories	Beverages, baked goods, confections, table-top sugar, substitutes, many processed foods	Sucrose (sugar), glucose, fructose, sorbitol, mannitol, corn syrup, high fructose corn syrup, saccharin, aspartame, sucralose, acesulfame potassium (acesulfame-K), neotame
Colour additives	Offset colour loss due to exposure to light, air, temperature extremes, moisture and storage conditions; correct natural variations in colour; enhance colours that	Many processed foods, (candies, snack foods margarine, cheese, soft drinks, jams/jellies, gelatins, pudding and pie fillings)	FD&C Blue Nos. 1 and 2, FD&C Green No. 3, FD&C Red Nos. 3 and 40, FD&C Yellow Nos. 5 and 6, Orange B, Citrus Red No. 2, annatto extract, beta-carotene, grape skin extract, cochineal

Contd.

Table 10.1: Various types of food additives/preservatives/ingredients
(Contd.)

Types of ingredients	What they do	Examples of uses	Names found on product labels
	occur naturally; provide colour to colourless and "fun" foods		extract or carmine, paprika oleoresin, caramel colour, fruit and vegetable juices, saffron (*Note*: Exempt colour additives are not required to be declared by name on labels but may be declared simply as colourings or colour added)
Flavours and spices	Add specific flavours (natural and synthetic)	Pudding and pie fillings, gelatin dessert mixes, cake mixes, salad dressings, candies, soft drinks, ice cream, BBQ sauce	Natural flavouring, artificial flavour, and spices
Flavour enhancers	Enhance flavours already present in foods (without providing their own separate flavour)	Many processed foods	Mono-sodium glutamate (MSG), hydrolyzed soy protein, auto-lyzed yeast extract, disodium guanylate or inosinate
Fat replacers (and components of formulations	Provide expected texture and a creamy "mouth-feel" in	Baked goods, dressings, frozen desserts, confections,	Olestra, cellulose gel, carrageenan, poly-dextrose, modified food

Contd.

Table 10.1: Various types of food additives/preservatives/ingredients
(Contd.)

Types of ingredients	What they do	Examples of uses	Names found on product labels
used to replace fats)	reduced-fat foods	cake and dessert mixes, dairy products	starch, micro-particulated egg white protein, guar gum, xanthan gum, whey protein concentrate
Nutrients	Replace vitamins and minerals lost in processing (enrichment), add nutrients that may be lacking in the diet (fortification)	Flour, breads, cereals, rice, macaroni, margarine, salt, milk, fruit beverages, energy bars, instant breakfast drinks	Thiamine hydrochloride, riboflavin (vitamin B_2), niacin, niacinamide, folate or folic acid, beta carotene, potassium iodide, iron or ferrous sulfate, alpha tocopherols, ascorbic acid, vitamin D, amino acids (L-trypto-phan, L-lysine, L-leucine, L-methionine)
Emulsifiers	Allow smooth mixing of ingredients, prevent separation Keep emulsified products stable, reduce stickiness, control crysta-llization, keep ingredients	Salad dressings, peanut butter, chocolate, margarine, frozen desserts	Soy lecithin, mono- and diglycerides, egg yolks, polysor-bates, sorbitan mono-stearate

Contd.

Table 10.1: Various types of food additives/preservatives/ingredients
(Contd.)

Types of ingredients	What they do	Examples of uses	Names found on product labels
	dispersed, and to help products dissolve more easily		
Stabilizers and thickeners, binders, texturizers	Produce uniform texture, improve "mouth-feel"	Frozen desserts, dairy products, cakes, pudding and gelatin mixes, dressings, jams and jellies, sauces	Gelatin, pectin, guar gum, carrageenan, xanthan gum, whey
pH control agents and acidulants	Control acidity and alkalinity, prevent spoilage	Beverages, frozen desserts, chocolate, low acid canned foods, baking powder	Lactic acid, citric acid, ammonium hydroxide, sodium carbonate
Leavening agents	Promote rising of baked goods	Breads and other baked goods	Baking soda, mono-calcium phosphate, calcium carbonate
Anti-caking agents	Keep powdered foods free-flowing, prevent moisture absorption	Salt, baking powder, confectioner's sugar	Calcium silicate, iron ammonium citrate, silicon dioxide
Humectants	Retain moisture	Shredded coconut, marsh-mallows, soft candies, confections	Glycerin, sorbitol
Yeast nutrients	Promote growth of yeast	Breads and other baked goods	Calcium sulfate, ammonium phosphate

Contd.

Table 10.1: Various types of food additives/preservatives/ingredients
(Contd.)

Types of ingredients	What they do	Examples of uses	Names found on product labels
Dough strengtheners and conditioners	Produce more stable dough	Breads and other baked goods	Ammonium sulfate, azodicarbonamide, L-cysteine
Firming agents	Maintain crispness and firmness	Processed fruits and vegetables	Calcium chloride, calcium lactate
Enzyme Preparations	Modify proteins, polysaccharides and fats	Cheese, dairy products, meat	Enzymes, lactase, papain, rennet, chymosin
Gases	Serve as propellant, aerate, or create carbonation	Oil cooking spray, whipped cream, carbonated beverages	Carbon dioxide, nitrous oxide

E numbers are codes for substances that are permitted to be used as food additives. Having a single unified list for food additives was first agreed upon in 1962 with food colouring. In 1964, the directives for preservatives were added, 1970 for antioxidants and 1974 for the emulsifiers, stabilisers, thickeners and gelling agents.

The numbering scheme follows that of the International Numbering System (INS) as determined by the Codex Alimentarius committee, though only a subset of the INS additives are approved for use in the European Union as food additives.

E number range	Subranges	Description
100–199 Colours	100–109	Yellow
	110–119	Orange
	120–129	Red
	130–139	Blue and violet
	140–149	Green
	150–159	Brown and black
	160–199	Gold and others

Contd.

E number range	Subranges	Description
200–299	200–209	Sorbate
Preservatives	210–219	Benzoate
	220–229	Sulphite
	230–239	Phenol and formate (methanoate)
	240–259	Nitrate
	260–269	Acetate (ethanoate)
	270–279	Lactate
	280–289	Propionate (propanoate)
	290–299	Others
300–399	300–305	Ascorbate (vitamin C)
Antioxidants and	306–309	Tocopherol (vitamin E)
acidity regulators	310–319	Gallate and erythorbate
	320–329	Lactate
	330–339	Citrate and tartrate
	340–349	Phosphates
	350–359	Malate and adipate
	360–369	Succinate and fumarate
	370–399	Others
400–499	400–409	Alginates
Thickeners,	410–419	Natural gums
stabilisers and	420–429	Other natural agents
emulsifiers	430–439	Polyoxyethene compounds
	440–449	Natural emulsifiers
	450–459	Phosphates
	460–469	Cellulose compounds
	470–489	Fatty acids and compounds
	490–499	Others
500–599 pH	500–509	Mineral acids and bases
regulators & anti-	510–519	Chlorides and sulphates
caking agents	520–529	Sulphates and hydroxides
	530–549	Alkali metal compounds
	550–559	Silicates
	570–579	Stearates and gluconates
	580–599	Others
600–699 Flavour	620–629	Glutamates and guanylates
enhancers	630–639	Inosinates
	640–649	Others
700–799	700–713	
Antibiotics		

Contd.

E number range	Subranges	Description
900–999	900–909	Waxes
Miscellaneous	910–919	Synthetic glazes
	920–929	Improving agents
	930–949	Packaging gases
	950–969	Sweeteners
	990–999	Foaming agents
1100–1599	1100–1599	New chemicals that do not fall into
Additional		standard classification schemes
chemicals		

Some important kinds of food additives are discussed below:

1. Acidulents

Confer sour or acid taste. Common acidulents include vinegar, citric acid, tartaric acid, malic acid, fumaric acid, and lactic acid.

2. Acidity Regulators

Acidity regulators are used for controlling the pH of foods for stability or to affect activity of enzymes. **Also called pH control agents**, these are food additives added to change or maintain pH (acidity or basicity).

They can be organic or mineral acids, bases, neutralizing agents, or buffering agents. Typical agents include these acids and their sodium salts: sorbic acid, acetic acid, benzoic acid, and propionic acid.

Acidity regulators are indicated by their E number, such as E260 (acetic acid), or simply listed as "food acid".

Acidity regulators differ from acidulants, which are often acidic but are added to confer sour flavours. They are not intended to stabilize the food, although that can be a collateral benefit.

3. Anticaking Agents

An anticaking agent is an additive placed in powdered or granulated materials, such as table salt or confectionaries to prevent the formation of lumps (caking) and for easing packaging, transport, and consumption. Anticaking agents keep powders such as milk powder from caking or sticking.

An anticaking agent in salt is denoted in the ingredients, for example, as "anti-caking agent (554)", which is sodium aluminosilicate, a man-made product. This product is present in many commercial table salts as well as dried milk, egg mixes, sugar products, and flours. In Europe, sodium ferrocyanide (535) and potassium ferrocyanide (536) are more common anticaking agents in table salt. "Natural" anticaking agents used in more expensive table salt include calcium carbonate and magnesium carbonate.

Some anticaking agents are soluble in water, others are soluble in alcohols or other organic solvents. They function either by absorbing excess moisture or by coating particles and making them water-repellent. Calcium silicate ($CaSiO_3$), a commonly used anti-caking agent, added to e.g. table salt, absorbs both water and oil.

The following anticaking agents are listed in order by their number in the Codex Alimentarius.

- o 341 tricalcium phosphate
- o 460(ii) powdered cellulose
- o 470b magnesium stearate
- o 500 sodium bicarbonate
- o 535 sodium ferrocyanide
- o 536 potassium ferrocyanide
- o 538 calcium ferrocyanide
- o 542 bone phosphate
- o 550 sodium silicate
- o 551 silicon dioxide
- o 552 calcium silicate
- o 553a magnesium trisilicate
- o 553b talcum powder
- o 554 sodium aluminosilicate
- o 555 potassium aluminium silicate
- o 556 calcium aluminosilicate
- o 558 bentonite
- o 559 aluminium silicate
- o 570 stearic acid
- o 900 polydimethylsiloxane

4. Antifoaming and Foaming Agents

Antifoaming agents reduce or prevent foaming in foods. Foaming agents do the reverse. An **anti-foaming agent** is a chemical additive that reduces and hinders the formation of foam in industrial process liquids.

Commonly used agents are insoluble oils, polydimethyl-siloxanes and other silicones, certain alcohols, stearates and glycols. The additive is used to prevent formation of foam or is added to break foam already formed.

When used as an ingredient in food, antifoaming agents are intended to curb effusion or effervescence in preparation or serving. The agents are included in a variety of foods such as chicken nuggets in the form of poly dimethyl siloxane (a type of silicone).

Silicone oil is also added to cooking oil to prevent foaming in deep-frying.

5. Antioxidants

Antioxidants such as vitamin C are preservatives by inhibiting the degradation of food by oxygen.

6. Bulking Agents

Bulking agents such as starch are additives that increase the bulk of a food without affecting its taste.

7. Food Colouring

Colourings are added to food to replace colours lost during preparation or to make food look more attractive. **These encompass** any dye, pigment or substance that imparts colour when it is added to food or drink. They come in many forms consisting of liquids, powders, gels, and pastes. It is used both in commercial food production and in domestic cooking.

People associate certain colours with certain flavours, and the colour of food can influence the perceived flavour. Colour additives are used in foods for many reasons including:

○ Offset colour loss due to exposure to light, air, temperature extremes, moisture and storage conditions
○ Correct natural variations in colour

o Enhance colours that occur naturally
o Provide colour to colourless and "fun" foods
o Make food more attractive and appetizing, and informative
o Allow consumers to identify products on sight, like candy flavours or medicine dosages

While naturally derived colours are not required to be certified by a number of regulatory bodies throughout the world, they still need to be approved for use in that country.

o Certified colours are synthetically produced and are used widely because they impart an intense, uniform colour, are less expensive, and blend more easily to create a variety of hues. Certified food colours generally do not add undesirable flavours to foods.

• Certified colour additives are categorized as either dyes or lakes.

■ **Dyes** dissolve in water and are manufactured as powders, granules, liquids or other special-purpose forms. They can be used in beverages, dry mixes, baked goods, confections, dairy products, pet foods and a variety of other products.

■ **Lakes** are the water insoluble form of the dye. Lakes are more stable than dyes and are ideal for colouring products containing fats and oils or items lacking sufficient moisture to dissolve dyes. Typical uses include coated tablets, cake and donut mixes, hard candies and chewing gums.

o Colours that are **exempt from certification** include pigments derived from natural sources such as vegetables, minerals or animals. Nature derived colour additives are typically more expensive than certified colours and may add unintended flavours to foods. Examples of exempt colours include annatto, beet extract, caramel, beta-carotene, turmeric and grape skin extract.

Artificial Colouring

As the 1900s began, a host of synthetic dyes and pigments became available. Originally, these were dubbed 'coal-tar' colours because the starting materials were obtained from bituminous coal. Many synthesized dyes were easier and less costly to produce and were superior in colouring properties when compared to naturally derived alternatives of the time.

In the US, the following seven artificial colourings are generally permitted in food as of 2016. The lakes of these colourings are also permitted except the lake of Red No. 3.

- o **FD&C Blue No. 1** : Brilliant Blue FCF, E133 (blue shade)
- o FD&C Blue No. 2: Indigotine, E132 (indigo shade)
- o FD&C Green No. 3: Fast Green FCF, E143 (turquoise shade)
- o FD&C Red No. 3: Erythrosine, E127 (pink shade, commonly used in glacé cherries)
- o **FD&C Red No. 40**: Allura Red AC, E129 (red shade)
- o **FD&C Yellow No. 5**: Tartrazine, E102 (yellow shade)
- o **FD&C Yellow No. 6**: Sunset Yellow FCF, E110 (orange shade)

Permitted for limited use in foods

Two dyes are allowed by the FDA for limited applications:
- o Citrus Red 2 (orange shade)—allowed only to colour orange peels.
- o Orange B (red shade)—allowed only for use in hot dog and sausage casings (not produced after 1978, but never delisted)

Delisted and Banned

- o FD&C Red No. 2 – Amaranth, E123
- o FD&C Red No. 4
- o FD&C Red No. 32 was used to colour Florida oranges.
- o FD&C Orange No. 1 was one of the first water-soluble dyes to be commercialized, and one of seven original food dyes allowed under the Pure Food and Drug Act of June 30, 1906.
- o FD&C Orange No. 2 was used to colour Florida oranges.

o FD&C Yellow No. 1, 2, 3, and 4

o FD&C Violet No. 1

Natural Food Dyes

Carotenoids (E160, E161, E164), chlorophyllin (E140, E141), anthocyanins (E163), and betanin (E162) comprise four main categories of plant pigments grown to colour food products. [28] Other colourants or specialized derivatives of these core groups include:

o Annatto (E160b), a reddish-orange dye, made from the seed of the achiote

o Caramel colouring (E150a-d), made from caramelized sugar

o Carmine (E120), a red dye, derived from the cochineal insect, *Dactylopius coccus*

o Elderberry juice

o Lycopene (E160d)

o Paprika (E160c)

o Turmeric (E100)

Blue colours are especially rare. One feasible blue dye currently in use is derived from spirulina.

FD&C Yellow No. 5, is used to colour beverages, dessert powders, candy, ice cream, custards and other foods. FDA's Committee on Hypersensitivity to Food Constituents concluded in 1986 that FD&C Yellow No. 5 might cause hives in fewer than one out of 10,000 people. It also concluded that there was no evidence the colour additive in food provokes asthma attacks. The law now requires Yellow No. 5 to be identified on the ingredient line. This allows the few who may be sensitive to the colour to avoid it.

8. Colour Retention Agents

In contrast to colourings, colour retention agents are used to preserve a food's existing colour. **Colour retention agents** are food additives that are added to food to prevent the colour from changing. Many of them work by absorbing or binding to oxygen before it can damage food (antioxidants). For example, ascorbic acid (vitamin C) is often added to brightly coloured fruits such as peaches during canning.

List of Colour Retention Agent

E number	Common name	Max permitted level	Sources	Application
E300	Ascorbic acid	GMP	Standard 1.3.1-Food Additives (Australian)	Wine, sparkling wine and fortified wine
		0.03% (w/w), or 0.02% (w/w) depending on the matrix	The Miscellaneous Food Additives Regulations 1995	Fruit and vegetable-based drinks, juices and baby foods
				Fat-containing cereal-based foods including biscuits and rusks

9. Emulsifiers

Emulsifiers allow water and oils to remain mixed together in an emulsion, as in mayonnaise, ice cream, and homogenized milk. An **emulsifier** (also known as an "emulgent") is a substance that stabilizes an emulsion by increasing its kinetic stability. One class of emulsifiers is known as "surface active agents", or surfactants.

Examples of food emulsifiers are:

- o Egg yolk—in which the main emulsifying agent is lecithin. In fact, *lecithos* is the Greek word for egg yolk.
- o Mustard—where a variety of chemicals in the mucilage surrounding the seed hull act as emulsifiers
- o Soy lecithin is another emulsifier and thickener

- ○ Pickering stabilization—uses particles under certain circumstances
- ○ Sodium phosphates
- ○ Sodium stearoyl lactylate
- ○ DATEM (Diacetyl Tartaric (Acid) Ester of Monoglyceride)—an emulsifier used primarily in baking

Oil-in-water emulsions are common in food products:

- ○ Crema (foam) in espresso—coffee oil in water (brewed coffee), unstable emulsion
- ○ Mayonnaise and Hollandaise sauce—these are oil-in-water emulsions that are stabilized with egg yolk lecithin, or with other types of food additives, such as sodium stearoyl lactylate
- ○ Homogenized milk—an emulsion of milk fat in water and milk proteins
- ○ Vinaigrette—an emulsion of vegetable oil in vinegar. If this is prepared using only oil and vinegar (i.e. without an emulsifier), an unstable emulsion results.

Water-in-oil emulsions are less common in food but still exist:

- ○ Butter—an emulsion of water in butter fat.

10. Flavours

Flavours are additives that give food a particular taste or smell, and may be derived from natural ingredients or created artificially.

Flavour is the sensory impression of food or other substance, and is determined primarily by the chemical senses of taste and smell. The "trigeminal senses", which detect chemical irritants in the mouth and throat as well as temperature and texture, are also important to the overall Gestalt of flavour perception. The flavour of the food, as such, can be altered with natural or artificial flavourants which affect these senses.

A *flavourant* is defined as a substance that gives another substance flavour, altering the characteristics of the solute, causing it to become sweet, sour, tangy, etc.

Of the three chemical senses, smell is the main determinant of a food item's flavour. While there are only five universally recognized basic tastes—sweet, sour, bitter, salty, and umami (savoury)—the number of food smells is unbounded. A food's flavour, therefore, can be easily altered by changing its smell while keeping its taste similar. This is exemplified in artificially flavoured jellies, soft drinks, and candies, which, while made of bases with a similar taste, have dramatically different flavours due to the use of different scents or fragrances. The flavourings of commercially produced food products are typically created by flavourists.

Flavourings are focused on altering the flavours of natural food product such as meats and vegetables, or creating flavour for food products that do not have the desired flavours such as candies and other snacks. Most types of flavourings are focused on scent and taste. A few commercial products exist to stimulate the trigeminal senses, since these are sharp, astringent, and typically unpleasant flavours.

There are three principal types of flavourings used in foods, under definitions agreed in the EU and Australia:

Type	Description
Natural flavouring substances	Flavouring substances obtained from plant or animal raw materials, by physical, microbiological or enzymatic processes. They can be either used in their natural state or processed for human consumption, but cannot contain any nature-identical or artificial flavouring substances.
Nature-identical flavouring substances	Flavouring substances that are obtained by synthesis or isolated through chemical processes, which are chemically and organoleptically identical to flavouring substances naturally present in products intended for human consumption. They cannot contain any artificial flavouring substances.
Artificial flavouring substances	Flavouring substances not identified in a natural product intended for human consumption, whether or not the product is processed. These are typically produced by fractional distillation and additional chemical manipulation of naturally sourced chemicals, crude oil or coal tar. Although they are chemically different, in sensory characteristics are the same as natural ones.

Chemical	Odour
Diacetyl, acetylpropionyl, acetoin	Buttery
Isoamyl acetate	Banana
Benzaldehyde	Bitter almond, cherry
Cinnamaldehyde	Cinnamon
Ethyl propionate	Fruity
Methyl anthranilate	Grape
Limonene	Orange
Ethyl decadienoate	Pear
Allyl hexanoate	Pineapple
Ethyl maltol	Sugar, cotton candy
Ethylvanillin	Vanilla
Methyl salicylate	Wintergreen

The compounds used to produce artificial flavours are almost identical to those that occur naturally. It has been suggested that artificial flavours may be safer to consume than natural flavours due to the standards of purity and mixture consistency that are enforced either by the company or by law.

While salt and sugar can technically be considered flavourants that enhance salty and sweet tastes, usually only compounds that enhance umami, as well as other secondary flavours are considered and referred to as *taste flavourants*. Artificial sweeteners are also technically flavourants.

Umami or "savoury" flavourants, more commonly called taste or flavour enhancers, are largely based on amino acids and nucleotides. These are typically used as sodium or calcium salts.

Umami flavourants recognized and approved by the European Union include:

Acid	Description
Glutamic acid salts	This amino acid's sodium salt, monosodium glutamate (MSG), is one of the most commonly used flavour enhancers in food processing. Mono and diglutamate salts are also commonly used.
Glycine salts	Simple amino acid salts typically combined with glutamic acid as flavour enhancers.

Contd.

Acid	Description
Guanylic acid salts	Nucleotide salts typically combined with glutamic acid as flavour enhancers.
Inosinic acid salts	Nucleotide salts created from the breakdown of AMP. Due to high costs of production, typically combined with glutamic acid as flavour enhancers.
5'-ribonucleotide salts	Nucleotide salts typically combined with other amino acids and nucleotide salts as flavour enhancers.

Certain organic and inorganic acids can be used to enhance sour tastes, but like salt and sugar these are usually not considered and regulated as flavourants under law. Each acid imparts a slightly different sour or tart taste that alters the flavour of a food.

Acid	Description
Acetic acid	Gives vinegar its sour taste and distinctive smell
Ascorbic acid	Found in oranges and green peppers and gives a crisp, slightly sour taste. Better known as vitamin C.
Citric acid	Found in citrus fruits and gives them their sour taste
Fumaric acid	Not found in fruits, used as a substitute for citric and tartaric acid
Lactic acid	Found in various milk or fermented products and give them a rich tartness
Malic acid	Found in apples and gives them their sour/tart taste
Phosphoric acid	Used in some cola drinks to give an acid taste
Tartaric acid	Found in grapes and wines and gives them a tart taste

11. Flour Treatment Agents

Flour treatment agents are added to flour to improve its colour or its use in baking. They are food additives combined with flour to improve baking functionality. Flour treatment agents are used to increase the speed of dough rising and to improve the strength and workability of the dough. While they are an important component of modern factory baking, some small-scale bakers reject them in favour of longer fermentation periods that produce greater depth of flavour. There are wide ranges of these conditioners used in factory baking, which fall into four main categories: Bleaching agents, oxidizing and reducing agents, enzymes, and emulsifiers. These agents are often sold as mixtures in a soy flour base, as only small amounts are required.

- o **Flour bleaching agents** are added to flour to make it appear whiter (freshly milled flour is yellowish), to oxidize the surfaces of the flour grains, and help with developing of gluten.
- o Various flour bleaching agents
  - azodicarbonamide (E927)
  - carbamide (E927b)
  - potassium bromate (E924, the component which gives bromated flour its name, used mainly in the US East and Midwest, acts as a bleaching agent, banned in some areas)
  - phosphates
  - malted barley
  - potassium iodate
- o **Oxidizing agents** are added to flour to help with gluten development. They may or may not also act as bleaching agents. Originally flour was naturally aged through exposure to the atmosphere. Oxidizing agents primarily affect sulfur-containing amino acids, ultimately helping to form disulfide bridges between the gluten molecules. The addition of these agents to flour will create a stronger dough.
  - Common oxidizing agents are:
    - *Ascorbic acid*: Ascorbic acid converts into its oxidizing form, dehydroascorbic acid (DHAA) during mixing.
- o **Reducing agents** help to weaken the flour by breaking the protein network. This will help with various aspects of handling a strong dough. The benefits of adding these agents are reduced mixing time, reduced dough elasticity, reduced proofing time, and improved machinability.
  - Common reducing agents are:

- L-cysteine (E920, E921; quantities in the tens of ppm range help soften the dough and thus reduce processing time)
- fumaric acid
- sodium bisulfate
- non-leavened yeast

o **Enzymes** are also used to improve processing characteristics. Yeast naturally produces both amylases and proteinases, but additional quantities may be added to produce faster and more complete reactions.

- Amylases break down the starch in flours into simple sugars, thereby letting yeast ferment quickly. Malt is a natural source of amylase.
- Proteases improve extensibility of the dough by degrading some of the gluten.
- Lipoxygenases oxidize the flour.

12. Glazing Agents

Glazing agents provide a shiny appearance or protective coating to foods. A glazing agent is a natural or synthetic substance that provides a waxy, homogeneous, coating to prevent water loss from a surface and provide other protection.

Natural Glazing Agents

Natural glazing agents have been found, usually in plants or insects. In nature, the agents are used to keep the moisture in the organism, but science has harnessed this characteristic by turning it into a glazing agent that acts as a coating. The glazing agent is made up of a substance that is classified as a wax. A natural wax is chemically defined as an ester with a very long hydrocarbon chain that also includes a long chain alcohol. However, in a wax there have been many different chemical structures that can be included in a definition of a wax, such as: Wax esters, sterol esters, ketones, aldehydes, alcohols, hydrocarbons, and sterols.

Examples are:
o Stearic acid (E570)
o Beeswax (E901)

- Candelilla wax (E902)
- Carnauba wax (E903)
- Shellac (E904)
- Microcrystalline wax (E905c), crystalline wax (E907)
- Lanolin (E913)
- Oxidized polyethylene wax (E914)
- Esters of colophonium (E915)
- Paraffin

Synthetic Glazing Agents

Science has produced glazing agents that mimic their natural counterparts. These components are added in different proportions to achieve the optimal glazing agent for a product.

Some of the characteristics that are looked for in all of the above industries are:

- *Preservation:* The glazing agent must protect the product from degradation and water loss. This characteristic can lead to a longer shelf life for a food.
- *Stability:* The glazing agent must maintain its integrity under pressure or heat.
- *Uniform viscosity:* This ensures a stronger protective coating that can be applied to the product as a homogeneous layer.
- *Industrial reproduction*: Because most glazing agents are used on commercial goods and therefore large quantities of glazing agent may be needed.

There are different variations of glazing agents, depending on the product, but they are all designed for the same purpose.

13. Humectants

Humectants prevent foods from drying out. These are hygroscopic substances used to keep things moist unlike desiccants which keep things dry. When used as a food additive, a humectant has the effect of keeping the foodstuff moist.

A humectant attracts and retains the moisture in the air nearby via absorption, drawing the water vapor into or beneath the organism's or object's surface.

While desiccants also attract ambient moisture, but adsorb—not absorb—it, by condensing the water vapor onto the surface, as a layer of film.

Examples of some humectants include:

o Aloe vera gel
o Alpha hydroxy acids such as lactic acid
o Egg yolk and egg white
o Glyceryl triacetate
o Honey
o Sugar alcohols (sugar polyols) such as glycerol, sorbitol, xylitol, maltitol
o Urea

Use in Food Industry

Some common humectants used in food are honey and glucose syrup both for their water absorption and sweet flavour. Glucose syrup also helps to retain the shape of the product better than other alternatives, for a longer period of time. In addition, some humectants are recognized in different countries as good food additives because of the increase in nutritional value that they provide, such as sodium hexametaphosphate.

Some of these humectants are seen in non-ionic polyols like sucrose, glycerin or glycerol and its triester (triacetin). These humectant food additives are used for the purpose of controlling viscosity and texture. Humectants also add bulk, retain moisture, reduce water activity, and improve softness. The main advantage of humectant food additives is that, since they are non-ionic, they are not expected to influence any variation of the pH aqueous systems.

Humectants are used in stabilization of food products and lengthening shelf life through food and moisture control. The available moisture determines microbial activity, physical properties, sensory properties and the rate of chemical changes, that if not controlled, are the cause of reduced shelf life.

Examples are dry cereal with semi-moist raisins, ice cream in a cone, chocolate, hard candy with liquid centers and cheese. Humectants are used to stabilize the moisture content of foodstuffs and are incorporated as food additives. Humectants are also used in military technology for the use of MREs and other military rations. A number of food items always need to be moist. The use of humectants reduces the available water, thus reducing bacterial activity. They are used for safety issues, for quality, and to have a longer shelf-life in food products.

An example of where humectants are used to keep food moist is in products like toothpaste as well as certain kinds of cookies. Regional kinds of cookies often use humectants as a binding agent in order to keep moisture locked into the center of the cookie rather than have it evaporate out. Humectants are favoured in food products because of their ability to keep consumable goods moist and increase shelf-life.

Tobacco Products

Humectants are used in the manufacturing of some tobacco products, such as cigarettes, e-cigarettes, and self-rolled tobacco. They are used to control and maintain the moisture content of the cut tobacco filler and add flavour.

The main health concern regarding e-cigarettes is that their production is not regulated, and there is immense uncertainty of quality control during manufacturing. Self-rolled tobacco contains more humectants, which are added to tobacco to make it taste better and keep from drying out. As the humectants burn, they unleash chemicals such as acreolein. Humectants are found in most cigarettes and are considered one of the most dangerous chemicals found in tobacco.

However, there have been conflicting claims about the degree to which these products warrant a health concern.

14. Tracer Gas

Tracer gas allow for package integrity testing to prevent foods from being exposed to atmosphere, thus guaranteeing shelf life.

15. Stabilizers

Stabilizers, thickeners and gelling agents, like agar or pectin give foods a firmer texture. While they are not true emulsifiers, they help to stabilize emulsions. Jam has an essential ingredient as jam.

16. Sweeteners

Sweeteners are added to foods for flavouring. Sweeteners other than sugar are added to keep the food calories low, or because they have beneficial effects for diabetes mellitus and tooth decay and diarrhoea.

17. Thickeners

Thickening agents are substances which, when added to the mixture, increase its viscosity without substantially modifying its other properties.

These are substances which can increase the viscosity of a liquid without substantially changing its other properties. Edible thickeners are commonly used to thicken sauces, soups, and puddings without altering their taste.

Some thickening agents are gelling agents (gellants), forming a gel, dissolving in the liquid phase as a colloid mixture that forms a weakly cohesive internal structure. Others act as mechanical thixotropic additives with discrete particles adhering or interlocking to resist strain.

Thickening agents can also be used when a medical condition such as dysphagia causes difficulty in swallowing. Thickened liquids play a vital role in reducing risk of aspiration for dysphagia patients.

Food thickeners frequently are based on either polysaccharides (starches, vegetable gums, and pectin), or proteins.

A flavourless powdered starch used for this purpose is a fecula (from the Latin faecula). This category includes starches as arrowroot, cornstarch, potato starch, sago, tapioca and their starch derivatives.

Vegetable gums used as food thickeners include alginin, guar gum, locust bean gum, and xanthan gum.

Proteins used as food thickeners include collagen, egg whites, furcellaran, and gelatin.

Sugars include agar and carrageenan.

Different thickeners may be more or less suitable in a given application, due to differences in taste, clarity, and their responses to chemical and physical conditions.

For instance:

- For acidic foods, arrowroot is a better choice than cornstarch, which loses thickening potency in acidic mixtures.
- At (acidic) pH levels below 4.5, guar gum has sharply reduced aqueous solubility, thus also reducing its thickening capability.
- If the food is to be frozen, tapioca or arrowroot is preferable over cornstarch, which becomes spongy when frozen.

Functional flours are produced from specific cereal variety (wheat, maize, rice or other) conjugated to specific heat treatment able to increase stability, consistency and general functionalities. These functional flours are resistance to industrial stresses such as acidic pH, sterilisation, freeze conditions, and can help food industries to formulate with natural ingredients. For the final consumer, these ingredients are more accepted because they are shown as "flour" in the ingredient list.

Flour is often used for thickening gravies, gumbos, and stews. It must be cooked in thoroughly to avoid the taste of uncooked flour. Roux, a mixture of flour and fat (usually butter) cooked into a paste, is used for gravies, sauces and stews.

- Cereal grains (oatmeal, couscous, farina, etc.) are used to thicken soups.
- Yogurt is popular in Eastern Europe and Middle East for thickening soups.
- Soups can also be thickened by adding grated starchy vegetables before cooking, though these will add their own flavour.
- Tomato puree also adds thickness as well as flavour.
- Egg yolks are a traditional sauce thickener in professional cooking; they have rich flavour and offer a velvety smooth texture but achieve the desired thickening effect only in a narrow temperature range.

Overheating easily ruins such a sauce, which can make egg yolk difficult to use as a thickener for amateur cooks. Other thickeners used by cooks are nuts (including rehan) or glaces made of meat or fish.

Many thickening agents require extra care in cooking. Some starches lose their thickening quality when cooked for too long or at too high a temperature; on the other hand, cooking starches too short or not hot enough might lead to an unpleasant starchy taste or cause water to seep out of the finished product after cooling. Also, higher viscosity causes foods to burn more easily during cooking. As an alternative to adding more thickener, recipes may call for reduction of the food's water content by lengthy simmering. When cooking, it is generally better to add thickener cautiously; if over-thickened, more water may be added but loss of flavour and texture may result.

Gelling agents are food additives used to thicken and stabilize various foods, like jellies, desserts and candies. The agents provide the foods with texture through formation of a gel. Some stabilizers and thickening agents are gelling agents.

Typical gelling agents include natural gums, starches, pectins, agar-agar and gelatin. Often they are based on polysaccharides or proteins.

Examples are:

o Alginic acid (E400), sodium alginate (E401), potassium alginate (E402), ammonium alginate (E403), calcium alginate (E404)—polysaccharides from brown algae

o Agar (E406, a polysaccharide obtained from red algaes)

o Carrageenan (E407, a polysaccharide obtained from red seaweeds)

o Locust bean gum (E410, a natural gum polysaccharide from the seeds of the Carob tree)

o Pectin (E440, a polysaccharide obtained from apple or citrus-fruit)

o Gelatin (E441, made by partial hydrolysis of animal collagen)

Commercial jellies used in East Asian cuisines include the glucomannan polysaccharide gum used to make "lychee cups" from the konjac plant, and aiyu or ice jelly from the *Ficus pumila* climbing fig plant.

Food thickening can be important for people facing medical issues with chewing or swallowing, as foods with a thicker consistency can reduce the chances of choking, or of inhalation of liquids or food particles, which can lead to aspiration pneumonia.

PRESERVATIVES

Preservatives prevent or inhibit spoilage of food due to fungi, bacteria and other microorganisms.

These are substances or chemicals which are added to products such as food and many other products to prevent decomposition by microbial growth or by undesirable chemical changes.

Broadly speaking preservation can be categorised as chemical and physical. Chemical preservation involves adding chemical compounds to the product.

Physical preservation involves use of physical processes such as UV-C radiation, freeze-drying, refrigeration or drying. Preservative food additives reduce the risk of food-borne infections, decrease microbial spoilage, and preserve fresh attributes and nutritional quality.

Antimicrobial Preservatives

Antimicrobial preservatives prevent degradation by bacteria. This method is the most traditional and ancient type of preserving—ancient methods such as pickling and adding honey prevent micro-organism growth by modifying the pH level.

The most commonly used antimicrobial preservative is lactic acid, nitrates and nitrites.

Some chemical preservatives are listed below:

E number	Chemical compound	Comment
E201 – E203	Benzoic acid, sodium benzoate	Used in acidic foods such as jams, salad dressing, juices, pickles, carbonated drinks, soy sauce
E214 – E219	Hydroxybenzoate and derivatives	Stable at a broad ph range
E270	Lactic acid	
E249 – E250	Nitrite	Used in meats to prevent botulism toxin

Contd.

E number	Chemical compound	Comment
E251 – E252	Nitrate	Used in meats
E280 – E283	Propionic acid and sodium propionate	Baked goods
E220 – E227	Sulfur dioxide and sulfites	Common for fruits
E200 – E203	Sorbic acid and sodium sorbate	Common for cheese, wine, baked goods

Antioxidants as Food Quality Preservatives

The free radical pathway is the first phase of the oxidative rancidification of fats. This process is slowed by antioxidants.

The oxidation process spoils most food, especially those with a high fat content. Fats quickly turn rancid when exposed to oxygen. Antioxidants prevent or inhibit the oxidation process. The most common antioxidant additives are ascorbic acid (vitamin C) and ascorbates. Thus, antioxidants are commonly added to oils, cheese, and chips.

Some important antioxidants used in food industry, as preservatives are listed below:

E number	Chemical compound	Comment
E300-304	Ascorbic acid, sodium ascorbate	Cheese, chips
E321	Butylated hydroxytoluene, butylated hydroxyanisole	Also used in food packaging
E310-312	Gallic acid and sodium gallate	Oxygen scavenger
E220–E227	Sulfur dioxide and sulfites	Beverages, wine
E306–E309	Tocopherols	Vitamin E activity

Natural Compounds for Food Preservation

Citric and ascorbic acids target enzymes that degrade fruits and vegetables, e.g. phenolase which turns surfaces of cut apples and potatoes brown. Ascorbic acid and tocopherol, which are vitamins, are common preservatives. Smoking entails exposing food to a variety of phenols, which are antioxidants. Natural preservatives include rosemary extract, hops, salt, sugar, vinegar, alcohol, diatomaceous earth and castor oil.

Chapter 11

Standardization of Recipes

An integral element of food product development or food service establishment is the standardization of the recipes of food products they cater to.

The United States Department of Agriculture (USDA) defines a standardized recipe as one that *"has been tried, adapted, and retried several times for use by a given foodservice operation and has been found to produce the same good results and yield every time when the exact procedures are used with the same type of equipment and the same quantity and quality of ingredients"*

To put it in simple terms *"A standardized recipe is a recipe that has been tried, tested, evaluated and adapted for use by a food service. It produces a consistent quality and yield every time when the exact procedures, equipment, and ingredients are used."*

A **standardized recipe** refers to a particular standard-of-use of certain metrics in cooking: Standard sizes, time, temperature, amount, etc. Abiding by this rule **creates uniformity** in kitchen produce, whether or not it is tangible or intangible.

In the kitchen, a standardized recipe is a crucial part of standardizing dishes, ingredients and elements in a restaurant that might lead to gain or loss during operational hours. Certain restaurants benchmark standardized recipes in their kitchen, some do not. There are pros and cons of using standardized recipes.

BENEFITS OF STANDARDIZED RECIPES

There are numerous benefits of recipe standardization which makes them indespensible for food service management programmes. Some are discussed below:

- o **Quality control:** Standardized recipes provide the same high-quality food every time they are used because they have been thoroughly tested and evaluated.
- o **Portion and yield control:** The amount of food that will be produced is the same every time with a standardized recipe, meaning it will reduce leftover food and make shortages much less likely.
- o **Cost control:** It is easier to manage buying and storing food when using the same ingredients in the same quantities every time a particular recipe is made. Standardized recipes provide consistent and accurate information for food cost control because the same ingredients and quantities of ingredients per serving are used each time the recipe is produced.
- o **Customer satisfaction:** When the research is done to find the types of foods customers want, recipe standardization ensures that they get the same quality, presentation and amount that they are accustomed to each time the meal is served. Well-developed recipes that appeal to customers are an important factor in maintaining and increasing sales. Standardized recipes provide this consistency and can result in increased customer satisfaction.
- o **Consistent nutrient content:** With so much testing and planning involved in a standardized recipe, one knows the exact nutritional content of the food every time it is prepared.
- o **Efficient purchasing procedures**: Purchasing is more efficient because the quantity of food needed for production is easily calculated from the information on each standardized recipe.
- o **Inventory control**: The use of standardized recipes provides predictable information on the quantity of food inventory that will be used each time the recipe is produced.

○ **Labor cost control**: Written standardized procedures in the recipe make efficient use of labor time and allow for planned scheduling of foodservice personnel for the work day. Training costs are reduced because new employees are provided specific instructions for preparation in each recipe.

○ **Increased employee confidence**: Employees feel more satisfied and confident in their jobs because standardized recipes eliminate guesswork, decrease the chances of producing poor food products, and prevent shortages of servings during meal service.

○ **Reduced record keeping**: A collection of standardized recipes for menu items will reduce the amount of information required on a daily food production record. Standardized recipes will include the ingredients and amounts of food used for a menu item. The food production record will only need to reference the recipe, number of planned servings, and leftover amounts.

Table 11.1: Pros and cons of standardization of recipes

Pros	Cons
• Creates an absolute standard in kitchen produce and cooking activities.	• Inconvenient due to referring each time or flipping pages of books or accessibility issues.
• Allows smooth transition between different kitchen staffs.	• *Time consuming*: During peak hours, a kitchen staff do not have time to waste, and every second counts.
• Maintains food quality and food standards during kitchen operational hours.	• *Better variations*: Some chefs prefer to follow their centric of taste, some are just worship their own believes.
• Guiding tool for newcomers to the kitchen.	• *Rules are meant to be broken*: There are always different people/consumers around your restaurant. When standardized recipes are not tested regularly

Contd.

Table 11.1: Pros and cons of standardization of recipes *(Contd.)*

Pros	Cons
	on the restaurant, inaccurate information may be provided in the standardized recipe.
• Refresh minds of kitchen staff after some time (eliminating guesswork).	• *A secret no more*: Some restaurateurs or chefs refrain from making a book of standardized recipe because they want to protect their food knowledge.
• Referral material should be there for any dispute.	• At certain times in a restaurant, a piece of recipe sheet can get lost.
• Base for costing when kitchen costs are calculated.	
• Be a great guide for implementing a new menu should there be any need.	
• Planning and costing purposes when a particular event needs accounting/kitchen control auditing.	
• Prevents raw food leftovers	

A recipe should consist of:

1. *Recipe title*: Name that adequately describes the recipe.
2. *Recipe category*: Recipe classification
3. *Ingredients*: Products used in a recipe.
4. *Weight/volume of each ingredient*: The quantity of each ingredient listed
5. *Preparation instructions (directions)*: Directions for preparing the recipe.
6. *Cooking temperature and time*: The cooking temperature and time, if appropriate.
7. *Serving size*: The amount of a single portion in volume and/or weight.
8. *Recipe yield*: The amount of product at the completion of production.
9. *Equipment and utensils to be used*: The cooking and serving equipment to be used.

Other possible components
10. Contribution to the food-based menu planning system

RECIPE WRITING STYLES

There are three common ways of writing a recipe:
1. Paragraph style recipes
2. List style recipes
3. Action style recipes

Paragraph Style Recipes

This way of writing a recipe is classic—and they serve their own purpose in writing that way. An example of a paragraph-style written recipe:

Put your skillet on the pan and turn on the heat to low. Now take a bowl, crack 2 fresh eggs inside and add in some salt and pepper. Next, grab a whisk and start beating it until it is mixed or quite fluffy. When your skillet is hot enough, add in 1 tbsp of oil, and swirl the oil around. You will notice the oil runs faster on hot pans. When your pan and oil is hot enough, turn on the heat to high and pour in your eggs. Leave the heat on high until your eggs (at the side of the pan) forms a solid texture. At this time, reduce your heat to low. When your egg is cooked enough, flip it over and top it off with some fresh herbs! Voilá!

Paragraph style recipes can work at certain extent. Be sure to choose your methods of writing well.

List Style Recipes

The list style writing of recipes is one of the easiest, practical and most common ways of writing a recipe. This method consists of two sections: The header and footer.

Header consists of different elements such as recipe title, temperature, yield, time, etc, while the footer contains methods to use these ingredients. An example of list style recipes:

Eggs with fresh herbs

2 Eggs, 1 tbsp oil, 1 tbsp fresh mixed herbs
1. Heat up your pan in low heat, crack two eggs into a bowl and add seasoning. Whisk well.

2. When your pan is hot enough, add in your oil and wait until it is hot.
3. Pour it in and turn your heat to high, until you see the sides of your eggs are actually solid in texture.
4. Reduce your heat to low, and cook the eggs well. Flip over.
5. Top it off with some fresh herbs and voilá!

Action Style Recipes

Action style recipes have been known as the killer way of listing recipes, amount, methods and ingredients in a very organized and well-mannered. The first step will usually contain ingredients and methods limited to only a particular food preparation, and the list continues and combines with step two and three. Here is an example (Table 11.2):

Table 11.2 : Example of action style recipe

Ingre-dients	Amount	Method	Misc info
Frying pan	1 skillet	1. Put on range and turn on heat.	Low heat, to heat up pan while you prep egss.
Eggs	2 no	1. Crack eggs into bowl.	Do not take too long—your pan is getting hotter every passing second.
Bowl	1 piece, medium	2. Stirr in seasoning and use whisk to mix well	Cook eggs under high heat until the sides show a solid texture.
Whisk		3. Pour your egg mixt-ure into the pan and cook.	
Season-ing		4. Flip your eggs over when the bottom is done.	
Fresh herbs	1 tbsp	5. Top your eggs off and serve.	Serve your eggs on a rice plate, please.

Action style recipes can be very directive and you can add in more information to your liking.

Choose which is best for you and your audience, then pick the right one and give them value.

Standard Elements in a Standardized Recipe

Although we may see certain standard recipe metrics in a standardized recipe that may be both relevant and irrelevant to you, there are certain practical usage to it, and customizing your standardized recipe a good way to go when you need to emphasize certain recipe metrics in a recipe sheet.

Common recipe elements in a standardized recipe

1. Ingredients
2. Temperature
3. Equipment and utensils needed
4. Amount
5. Method
6. Media (picture/video)

It is always a good idea to include explanations within each head to better explain the elements.

Recommended Standard Recipe Elements to Add

These recommended standard recipe elements are absolutely optional and should only be included at selected times. Note that most recipes require only the simplest of steps to take, and portrayal of information should be as concise, clear and to the point as possible.

1. **Taste:** At what degree should this dish taste like, and how you can stretch its seasoning properties from there.
2. **Precautions and warnings:** Precautions while handling these food mix or cooking methods.
3. **Tips and advice:** Best way to beef up preparation methods and cook well without the need for practical training.
4. **What to do while waiting:** Important steps or methods to follow or take while waiting cooking or preparing a food ingredient or food ingredient mixes, etc.
5. **Alternatives:** Alternatives to this cooking method, or that food ingredient which might not be available in certain areas of the world. If there is any alternative way to do it, it should be pointed out.

6. **Garnishing recommendations:** This should be included and portrayed after recipe methods.

Phases of Recipe Standardization

The recipe standardization process can be summarized in three phases:

1. **Recipe verification:** Recipe verification consists of reviewing the recipe in detail, preparing it, verifying its yield, and recording changes.
2. **Product evaluation:** Product evaluation focuses on determining the acceptability of the product produced from the recipe.
3. **Quantity adjustment:** Changing the recipe yield and ingredient amounts occurs in the quantity adjustment phase.

A recipe may go through these phases several times before becoming standardized at the necessary quantity for an operation.

Decisions made during each phase determine the flow of a recipe through this recipe standardization process.

Once a recipe has been standardized for an operation, the standardization process should not have to be repeated unless changes occur in availability of ingredients or equipment.

1. **Recipe verification phase:** The first phase of the recipe standardization process is the recipe verification phase. This phase includes four major processes: Review the recipe, prepare the recipe, verify the recipe yield, and record changes to the recipe.

   a. *Review the recipe:* Begin by working on only one recipe at a time. Review the recipe to be standardized. Look to see if the recipe contains the following information and review each one with a clear focus.

   1. Recipe title
   2. Recipe category
   3. Ingredients
   4. Weight/volume for each ingredient
   5. Preparation instructions (directions)

6. Cooking temperature and time, if appropriate
7. Serving size
8. Recipe yield
9. Equipment and utensils to be used

Reviewing the recipe for this information must be done before preparing the recipe.

b. *Prepare the recipe*: Once the recipe has been reviewed, it can be prepared. Throughout the process of making the recipe, keep careful notes about any variations. Record this information directly on the recipe for future reference. Cooking time to reach the internal temperature and product quality may vary slightly depending on the type and age of equipment.

c. *Verify yields*: "Verify yields" includes verifying ingredient, recipe, and serving yields. Yields can vary depending on factors such as product quality, preparation techniques, and cooking times and temperatures.

Verification of the recipe yield occurs once all of the ingredients have been combined and the recipe preparation completed. The yield can be determined several ways depending on the recipe.

d. *Record changes:* Notes of any changes or concerns should be recorded on the recipe during the verification phase.

2. **Product evaluation phase:** Product evaluation follows the recipe verification phase and is an important part of the recipe standardization process. It will help determine acceptability of the recipe and will provide objective information that can be used to further improve the recipe. Recipe evaluation should include the manager, foodservice staff members, and customers.

Two types of evaluation occur in the evaluation phase: Informal and formal.

a. *Informal evaluation:* Informal evaluation involves only the foodservice managers and employees.

During informal evaluation, the product is prepared for the first time in the operation and an assessment is made of whether efforts to standardize the recipe should continue.

Three decisions are possible as a result of the informal evaluation of a recipe.

i. First, if the product was found to be totally unacceptable based on several of the informal evaluation criteria, the decision may be made to discontinue any further work on standardizing the recipe.

ii. If most of the informal evaluation criteria were rated as acceptable, the recipe may go back to the verification phase to allow for changes to be made to the recipe and a new version of the recipe prepared.

iii. Finally, if all evaluation criteria were rated as acceptable in the informal evaluation, then the recipe may be prepared for formal evaluation.

b. *Formal evaluation*: Formal evaluation occurs when the foodservice staff believes a recipe has potential for service in their operation. Procedures for conducting a formal evaluation of the recipe include:

i. *Select a group(s) of people to taste the sample recipe.* Usually 10 or fewer people should sample a food item at a given time.

ii. *Choose an evaluation form.* The evaluation form used should be appropriate for the age of the group members who are sampling the food items.

iii. *Prepare the sample recipe.* Once a group has been selected to sample the product(s) and an evaluation form has been selected, the recipe can be prepared for evaluation. Typically, recipes for sampling are made in small quantities such as for servings of 25.

iv. *Set up the sampling area.* The area to be used for sampling should be prepared with drinking water, eating and serving utensils, napkins, evaluation forms, and pens or pencils. If more than one food item is being evaluated, evaluators should be provided with unsalted soda crackers to nibble between foods. The cracker will help prevent flavour carryover from the first food.

v. *Have participants taste and evaluate the food.* Tasting procedures should be explained to those who will be evaluating the product, and the evaluation form should be reviewed with them prior to tasting. Remind evaluators of the importance of not making verbal comments about the food during the tasting. If asking for an evaluation of qualities such as moistness and/or temperature, explain what these terms mean.

vi. *Summarize the results.* The evaluation form used will help determine the way results are summarized.

vii. *Determine future plans for the recipe based on evaluation results.* Based on the formal evaluation results, the recipe will be accepted as is, rejected, or changed. If the formal evaluation comments are positive and the recipe is accepted as is, no further changes in ingredients will be needed. At this point a decision is made on whether the recipe is in the correct quantity or not.

If a different yield is needed, the recipe moves to the quantity adjustment phase of the recipe standardization process.

If no additional quantity adjustment is needed, the recipe is considered standardized.

If the evaluation comments are very poor, the recipe likely will be rejected and no further work will be done to standardize it for an operation.

If the evaluation comments were neither very good nor very poor, additional work on the recipe may be needed. This likely would mean that the recipe would go back through the verification phase with changes being made to ingredients, preparation instructions, or cooking procedures.

3. **Quantity adjustment phase:** When a recipe has been evaluated positively in the evaluation phase but is not in the desired quantity, it would move to the quantity adjustment phase of recipe standardization. There are several methods that can be used to adjust a recipe to get to the desired number of servings (yield). Some methods are done manually; others involve use of the computer.

The commonest manual method used is the Factor Method. It essentially comprises three steps:

1. **Calculate the factor:** The factor is a multiplier that will be used to increase or decrease the quantity of ingredients in a recipe. The factor is determined by dividing the desired yield (in number of servings) by the current recipe yield (in number of servings).

    Desired yield ÷ Current yield = Factor

    For example, if a manager wishes to make 250 servings and the current recipe produces 100 servings, divide 250 by 100; the factor would be 2.5.

    $$250 \div 100 = 2.5$$

2. **Multiply ingredients with the factor:** Each ingredient quantity in a recipe is multiplied by the factor to determine the ingredient quantity needed to produce the new yield. Ingredient quantities given as fractions would need to be converted to decimals prior to doing this calculation.

3. **Change amounts into more common measurements.** Often, the result of the mathematical calculations is a quantity that is hard to measure or not commonly used. These quantities may need to be converted to a more common measurement. Rounding to the nearest common measure also may occur.

Recipe Standardization Steps

1. Prepare a recipe to be standardized and test it until a high-quality product is produced that is acceptable. This step of the standardization process should include sensory evaluation. In addition, evaluate case of preparation and time commitment to prepare the recipe.

Table 11.3: Comparison of standardized recipe adjustment methods

Methods	Advantages	Disadvantages	Initial Recipe	Final Recipe
Factor method	• Can be used for any recipe • Easy to use	• Math skills required	• Can start with any recipe and desired yield	• Final recipe can yield any number • of servings desired
Direct reading tables method	• Minimal math skills needed	• Direct reading tables must be available	• Must have yield of 25 servings or multiples of 25 servings	• Yield of 25 servings or multiples of 25 servings (i.e. 200, 175, 500)
Percentage method	• Further adjustments to a single recipe are easy after initial ingredient percentages are calculated	• Many steps in process • Math skills required • Must use weights for all ingredients • Must calculate and adjust for handling loss	• Can start with any recipe and yield • Initial recipe ingredients must be weights	• Yield can be very amount desired • All final ingredients are in weighs
Computerised recipe adjustment	• Adjustments easy after recipe entered on computer • No math skills needed	• Computer programs can be expensive • Some programs require ingredients to	• Can start with any recipe and yield	• Final recipe can yield any number of servings desired

Contd.

Table 11.3: Comparison of standardized recipe adjustment methods
(Contd.)

Methods	Advantages	Disadvantages	Initial Recipe	Final Recipe
		be entered in weights only		
		• Ingredient quantities may be listed in decimals		

2. Determine portion size if that information is not available from the quantity recipe that is being standardized for the food service operation.

3. Calculate what a portion contributes to the meal pattern and make adjustments, where needed.

4. Retest the recipe if any changes were made.

5. Develop a written recipe that includes:

   a. Name of recipe (reflects contents and appeals to customers.

   b. Number/Category /Meal type for easy access.

   c. Exact ingredients by form to use (canned, frozen, dehydrated) and any pre-preparation steps needed (diced, chopped, grated).

   d. Detailed step-by-step procedures for preparation, cooking and serving. Include all steps for assembling ingredients.

   e. Cooking temperatures, cook time, and holding temperatures.

   f. Portion sizes(s) for single serving.

   g. Total recipe yield (measured or weighed), pan size, number of pans (if more than one), weight or measure in a pan.

   h. Equipment and specific serving utensil(s).

Chapter 12

Food Packaging, Graphics and Labelling

In today's society, packaging is pervasive and essential. It surrounds, enhances and protects the goods one buys, from processing and manufacturing, through handling and storage, to the final consumer. Without packaging, materials handling would be a messy, inefficient and costly exercise and modern consumer marketing would be virtually impossible.

The highly sophisticated packaging industries which characterize modern societies today are far removed from the simple packaging activities of earlier times. Food packaging lies at the very heart of the modern food industry and successful food packaging technologists emerge from a wide-ranging background drawn from a multitude of disciplines.

Despite the important role packaging plays, it is often regarded as a necessary evil or an unnecessary cost. Furthermore, in the view of many consumers, packaging is, at best, somewhat superfluous, and, at worst, a serious waste of resources and an environmental menace. Such views arise because the functions which packaging has to perform are either unknown or not considered in full. By the time most consumers come into contact with a package its job, in many cases, is almost over, and it is perhaps understandable that the view that excessive packaging has been used gains some credence.

Packaging has been defined as a socio-scientific discipline which operates in society, to ensure delivery of goods to the ultimate consumer of those goods in the best condition intended for their use.

The Packaging Institute International defines packaging as the enclosure of products, items or packages in a wrapped pouch, bag, box, cup, tray, can, tube, bottle or other container form to perform one or more of the following functions: Containment, protection, preservation, communication, utility and performance.

If the device or container performs one or more of these functions, it is considered a package.

Other definitions of packaging include a co-ordinated system of preparing goods for transport, distribution, storage, retailing and end-use, a means of ensuring safe delivery to the ultimate consumer in sound condition at optimum cost, and a techno-commercial function aimed at optimizing the costs of delivery while maximizing sales.

The package is the physical entity that contains the product.

The various levels of package are:

- A **primary package** is one which is in direct contact with the contained product. It provides the initial, and usually the major, protective barrier. Examples of primary packages include metal cans, paperboard cartons, glass bottles and plastic pouches. It is frequently only the primary package which the consumer purchases at retail outlets.

- A **secondary package** contains a number of primary packages, for example, a corrugated case or box. It is the physical distribution carrier and is sometimes designed so that it can be used in retail outlets for the display of primary packages.

- A **tertiary package** is made up of a number of secondary packages, the most common example being a stretch-wrapped pallet of corrugated cases. In interstate and international trade, a quaternary package is frequently used to facilitate the handling of tertiary packages. This is generally a metal container up to 40 m in length which can hold many pallets and is intermodal in nature, that is, it can be transferred to or from ships, trains, and flatbed trucks by giant cranes. Certain containers are also able to have their temperature, humidity and gas atmosphere controlled; this is necessary in particular situations such as the transportation of frozen foods, chilled meats and fresh fruits and vegetables.

FUNCTIONS OF PACKAGING

A package provides protection, tampering resistance, and special physical, chemical, or biological needs. It may bear a nutrition facts label and other information about food being offered for sale.

○ It protects contents from contamination and spoilage, makes it easier to transport and store goods and provides uniform measures of contents.

○ By allowing brands to be created and standardized, it makes advertising meaningful and large-scale distribution possible.

○ Special kinds of packages with dispensing caps, sprays and other convenience features make products more usable.

○ Packages serve as symbols of their contents and a way of life and, just as they can very powerfully communicate the satisfaction a product offers, they are equally potent symbols of wastefulness once the product is gone.

A package must do four important things:
○ It must protect the contents
○ Promote the product
○ Inform the consumer
○ It must be convenient for the consumer.

The functions of a food package were defined by the Codex Alimentarius Commission in 1985 as follows:

"Food is packaged to preserve its quality and freshness, add appeal to consumers and to facilitate storage and distribution."

Packaging and package labelling have several objectives
1. **Protection:** This is often regarded as the primary function of the package: To protect its contents from outside environmental effects, such as water, moisture vapour, gases, odors, micro-organisms, dust, shocks,

vibrations and compressive forces, and to protect the environment from the product.

For the majority of food products, the protection afforded by the package is an essential part of the preservation process. For example, aseptically packaged milk and fruit juices in paperboard cartons only remain aseptic for as long as the package provides protection. Likewise, vacuum-packaged meat will not achieve its desired shelf life if the package permits oxygen to enter. In general, once the integrity of the package is breached the product is no longer preserved.

a. **Physical protection**: The food enclosed in the package may require protection from, among other things, shock, vibration, compression, temperature, bacteria, etc.

b. **Barrier protection**: A barrier from oxygen, water vapour, dust, etc., is often required. Permeation is a critical factor in design. Some packages contain desiccants or oxygen absorbers to help extend shelf life. Modified atmospheres or controlled atmospheres are also maintained in some food packages. Keeping the contents clean, fresh, and safe for the intended shelf life is a primary function.

2. **Containment or agglomeration**: Small items are typically grouped together in one package to allow efficient handling. Liquids, powders, and granular materials need containment. The "package", whether it is a bottle of cola or a bulk cement rail wagon, must contain the product to function successfully. Without containment, product loss and pollution would be widespread. The containment function of packaging makes a huge contribution for protecting the environment from the myriad of products which are moved from one place to another on numerous occasions each day in any modern society. Faulty packaging (or under-packaging) could result in major pollution of the environment.

3. **Communication:** There is an old saying that "a package must protect what it sells and sell what it protects".

The modern methods of consumer marketing would fail were it not for the messages communicated by the package." The ability of consumers to instantly recognize products through distinctive branding and labelling enables supermarkets to function on a self-service basis.

  a. **Information transmission**: Packages and labels communicate how to use, transport, recycle, or dispose of the package or product. Some types of information are required by governments.

  b. **Marketing**: The packaging and labels can be used by marketers to encourage potential buyers to purchase the product. Package design has been an important and constantly evolving phenomenon for several decades. Marketing communications and graphic design are applied to the surface of the package and (in many cases) the point of sale display.

4. **Security**: Packaging can play an important role in reducing the security risks of shipment. Packages can be made with improved tamper resistance to deter tampering and also can have tamper-evident features to help indicate tampering. Packages can be engineered to help reduce the risks of package pilferage; some package constructions are more resistant to pilferage and some have pilfer-indicating seals. Packages may include authentication seals to help indicate that the package and contents are not counterfeit. Packages also can include anti-theft devices, such as dye packs, RFID tags, or electronic article surveillance tags, that can be activated or detected by devices at exit points and require specialized tools to deactivate. Using packaging in this way is a means of retail loss prevention.

5. **Convenience**: Packages can have features which add convenience in distribution, handling, stacking, display, sale, opening, reclosing, use, and reuse. Modernization and industrialization have precipitated tremendous changes in lifestyles and the packaging

industry has had to respond to those changes. Now an ever-increasing number of households are single-person, many couples either delay having children or opt not to at all and a greater percentage of women are in the workforce than ever before.

All these changes, as well as other factors such as the trend towards the demand for a wide variety of food and drink at outdoor functions such as sports events and increased leisure time, have created a demand for greater convenience in household products; products designed around principles of convenience include foods which are pre-prepared and can be cooked or reheated in a very short time, preferably without removing them from their primary package, and sauces, dressings and condiments that can be applied simply through aerosol or pump-action packages which minimize mess. Thus packaging plays an important role in meeting the demands of consumers for convenience.

The package functions by reducing the output from industrial production to a manageable, desirable "consumer" size.

An associated aspect is the shape (relative proportions) of the primary package with regard to consumer convenience (e.g. easy to hold, open and pour as appropriate) and efficiency in building into secondary and tertiary packages.

6. **Portion control**: Single-serving packaging has a precise amount of contents to control usage. Bulk commodities (such as salt) can be divided into packages that are a more suitable size for individual households. It also aids the control of inventory: Selling sealed one-liter bottles of milk, rather than having people bring their own bottles to fill themselves.

PACKAGE INTERACTIONS

The packaging has to perform its functions in three different environments.

Physical Environment

This is the environment in which physical damage can be caused to the product. It includes shocks from drops, falls and bumps, damage from vibrations arising from transportation modes including road, rail, sea and air and compression and crushing damage arising from stacking during transportation or storage in warehouses, retail outlets and the home environment.

Ambient Environment

This is the environment which surrounds the package. Damage to the product can be caused as a result of gases (particularly 02), water and water vapour, light (particularly UV radiation) and temperature, as well as micro-organisms (bacteria, fungi, molds, yeasts and viruses) and macro-organisms (rodents, insects, mites and birds) which are ubiquitous in many warehouses and retail outlets.

Contaminants in the ambient environment such as exhaust fumes from automobiles and dust and dirt can also find their way into the product unless the package acts as an effective barrier.

Human Environment

This is the environment in which the package interacts with people and designing packages for this environment requires knowledge of the variability of consumers' capabilities including vision, strength, weakness, dexterity, memory and cognitive behaviour. It includes knowledge of the results of human activity, such as liability, litigation, legislation and regulation. Since one of the functions of the package is to communicate, it is important that the messages are clearly received by consumers. In addition, the package must contain information required by law such as nutritional content and net weight.

Important points for convenient packages are:
o The package should be simple to hold, open and use.
o If not entirely consumed when the package is first opened, the package should be re-sealable and retain the quality of the product until completely used.

○ The package should contain a portion size which is convenient for the intended consumers; a package which contains too much product that it deteriorates before being completely consumed clearly contains too large a portion.

SELECTION CRITERIA FOR PACKAGES

A number of criteria must be considered when selecting a packaging system for a food. These include:

1. The stability of the food with respect to the deteriorative chemical, biochemical and microbiological reactions which can occur. The rates of these reactions depend on both intrinsic (compositional) and extrinsic (environmental) factors.

2. The environmental conditions to which the food will be exposed during distribution and storage. The ambient temperature and humidity are the two most important environmental factors and they dictate the barrier properties required of the package.

3. The compatibility of the package with the method of preservation selected. For example, if the food is being thermally processed after packing, then the packaging must obviously be able to withstand the thermal process. Likewise, if the food is to be stored at freezer temperatures after packing, then the packaging must be able to perform at these temperatures.

4. The nature and composition of the specific packaging material and its potential effect on the intrinsic quality and safety of the packaged food as a consequence of the migration of components from the packaging material into the food.

Putting it in a nutshell an ideal packaging material should have the following qualities:

○ It should contain the content within it
○ It should not affect the flavour of the product packaged it
○ Stable performance over large range of temperature.
○ Adequate compulsive strength and sufficient impact and puncture strength.

- o Sufficient thickness of cushioning materials with sufficient ventilation, space for rapid cooling of product.
- o Protect the product from O_2, moisture and light.
- o Protection of the content form adulterations.
- o Closure characteristics such as opening, sealing, resealing and pouring.
- o Low cost and availability.
- o It should be non-toxic in nature.
- o Proper labelling, storing, marketing appeal to sale, etc.

Migration

In food packaging terminology, *migration* is generally used to describe the transfer of substances from the package to the food. Substances that are transferred to the food as a result of contact or interaction between the food and the packaging material are often referred to as *migrants*.

However, it is important to note that migration is a two-way process because constituents of the food can also migrate into the packaging material.

In addition, compounds present in the environment surrounding the packaged food can be absorbed by the packaging and migrate into the food. For example, perfumes from soaps can be picked up by fatty foods under certain circumstances, which depend, among other factors, on the nature of the packaging materials used for the soap and the food, as well as on the proximity of the two products and the time of exposure.

It is important to distinguish between *overall migration* (OM; originally referred to as *global migration*) and *specific migration* (SM).

OM is the sum of all (usually unknown) mobile packaging components released per unit area of packaging material under defined test conditions. OM is therefore a measure of all compounds transferred into the food whether they are of toxicological interest or not, and will include substances that are physiologically harmless.

SM relates to an individual and identifiable compound only.

Different types of packaging materials and their suitability for food products

Table 12.1: Packaging type

Aseptic processing	Primary	Liquid whole eggs or dairy products
Trays	Primary	Portion of fish or meat
Bags	Primary	Potato chips, apples, rice
Boxes	Secondary	Corrugated box of primary packages: Box of cereal cartons, frozen pizzas
Cans	Primary	Can of tomato soup
Cartons, coated paper	Primary	Carton of eggs, milk or juice cartons
Flexible packaging	Primary	Bagged salad
Pallets	Tertiary	A series of boxes on a single pallet used to transport from the manufacturing plant to a distribution center
Wrappers	Tertiary	Used to wrap the boxes on the pallet for transport

Types of Packaging Materials

There are a number of materials used for packaging. Some of the important ones are discussed below:

1. Paper
2. Glass
3. Plastics
4. Metal
5. Edible films
6. Retortable pouches and trays
7. Cloth materials
8. Wooden containers
9. Composite containers
10. Regenerated cellulose
11. Cellulose acetate

1. Paper

Pulp is the raw material for the production of paper, paperboard, corrugated board and similar manufactured products. It is obtained from plant fiber and is therefore a renewable resource. There are three main constituents of wood cell wall:

Cellulose: This is a long chain, linear polymer built-up of a large numbers of glucose molecules and is the most abundant, naturally occurring organic compound. Cellulose is moderately resistant to the action of chlorine and dilute sodium hydroxide under mild conditions, but is modified or dissolved under more severe conditions. It is relatively resistant to oxidation and therefore bleaching operations can be used to remove small amounts of impurities such as lignin without appreciable damage to the strength of the pulp.

Hemicelluloses: These are lower molecular weight mixed sugar polysaccharides consisting of one or more of the following molecules: Xylose, mannose, arabiose, and galactose. Hemicelluloses are usually soluble in dilute alkalis.

Lignin: This is highly branched, thermoplastic polymer of uncertain size, built up largely from substituted phenyl-propane units. It has no fiber forming properties and is attacked by chlorine and sodium hydroxide with formation of soluble, dark brown derivatives. It softens at about 160°C.

It is generally used for the following:
1. Paper bags, wrapping, packaging papers and infusible tissues, e.g. tea and coffee bags, sachets, pouches, overwrapping paper, sugar and flour bags, carrier bags
2. Multiwall paper sacks
3. Folding cartons and rigid boxes
4. Corrugated and solid fiberboard boxes (shipping cases)
5. Paper-based tubes, tubs and composite containers
6. Fire drums
7. Liquid packaging
8. Moulded pulp containers
9. Labels
10. Sealing tapes
11. Cushioning materials
12. Cap liners (sealing wads) and diaphragms (membranes).

Paper and paperboard packaging is used over a wide temperature range, from frozen food storage to the high temperatures of boiling water and heating in microwave and conventional radiant heat ovens.

Paper and paperboard, however, can acquire barrier properties and extended functional performance, such as heat sealability for leak-proof liquid packaging, through coating and lamination with plastics, such as polyethylene (PE), polypropylene (PP), polyethylene terephthalate (PET or PETE) and ethylene vinyl alcohol (EVOH), and with aluminium foil, wax, and other treatments. Packaging made solely from paperboard can provide a wide range of barrier properties by being overwrapped with a heat sealable plastic film such as polyvinylidene chloride (PVdC) coated oriented polypropylene (OPP or BOPP).

The features of paper and paperboard which make these materials suitable for packaging relate to appearance and performance. These features are determined by the type of paper and paperboard—the raw materials used and the way they have been processed.

- o **Appearance:** Appearance relates to the visual impact of the pack and can be expressed in terms of colour, smoothness and whether the surface has a high or low gloss (matte) finish. Colour depends on the choice of fiber for the outer surface, and also, where appropriates, the reverse side. As described above, the choice is white, brown or grey. In addition some liners for corrugated board comprise a mix of bleached and brown fibers. Other colours are technically possible either by using fibers dyed to a specific colour or coated with a mineral pigment coloured coating.

- o **Performance:** Performance properties are related to the level of efficiency achieved during the manufacture of the pack, in printing, cutting and creasing, gluing and the packing operation. Performance properties are also related to pack compression strength in storage, distribution, at the point of sale and in consumer use. Specific measurable properties include stiffness, short span compression (rigidity) strength, tensile strength, wet strength, % stretch, tear strength, fold endurance, puncture resistance and ply bond strength. Other performance properties relate to moisture content, air permeability, water absorbency, surface friction, surface tension, ink absorbency, etc. Chemical properties include pH, whilst chloride and sulphate residues are relevant for aluminium foil lamination. Flatness is easily evaluated but is a complicated issue as lack of flatness can arise from several potential causes, from the hygro-sensitivity characteristics of the fiber, manufacturing variables and handling at any stage including printing and use. Neutrality with respect to odor and taint, and product safety are performance needs which are important in the context of paper and board packaging which is in direct or close proximity to food.

Types of Paper

Paper is divided into two broad categories: Fine papers, generally made of bleached pulp, and typically used for writing paper, bond, book and cover papers, and coarse papers, generally made of unbleached Kraft softwood pulps and used for packaging. Main types of packaging papers are:

- **Kraft paper:** This is typically a coarse paper with exceptional strength, often made on a fourdrinier machine and then either machine—glazed on a Yankee dryer or machine.
- **Bleached paper:** These are manufactured from pulps which are relatively white, bright and soft and receptive to the special chemicals necessary to develop many functional properties. They are generally more expensive and weaker than unbleached papers. Their aesthetic appeal is frequently augmented by day coating on one or both sides.
- **Greaseproof paper:** This is a translucent, machine finished paper which has been hydrated to give oil and grease resistance. Prolonged beating or mechanical refining is used to break the cellulose fibers which absorb so much water that they become superficially gelatinized and sticky.

- ○ **Glassine paper:** Glassine paper derives its name from its glassy, smooth surface, high density and transparency. It is produced by further treating greaseproof paper in a super calendar.
- ○ **Vegetable parchment:** Vegetable parchment takes its name from its physical similarity to animal parchment, which is made from animal skins. Because of its grease resistance and wet strength, it strips away easily from food material without defibering, thus finding use as an interleaver between slices of food such as meat or pastry. It was first used for wrapping fatty foods such as butter.
- ○ **Tissue paper:** Tissue papers range from semitransparent to totally opaque, and can be waxed. They are generally either machine-finished (MF) or machine-glazed (MG). MG papers may also be machine-finished to improve the smoothness on both sides.
- ○ **Waxed paper:** Waxed papers provide a barrier against penetration of liquids and vapours. Wet waxed papers have a continuous surface film on one or both sides achieved by shock-chilling the waxed web immediately after application of the wax. This also imparts a high degree of gloss on the coated surface. Dry waxed papers are produced using heated rolls and do not have a continuous film on the surfaces. Wax-laminated papers are bonded with a continuous film of wax which acts as an adhesive. The primary purpose of the wax is to provide a moisture barrier and a heat sealable laminate.

Types of Paper Boards

Paperboards are made from the same raw materials as papers. They normally are made on the cylinder machine and consist of two or more layers of different quality pulps. The types of paperboard used in food packaging include:

- ○ **Chipboard**: Chipboard is made from a mixture of repulped waste with chemical and mechanical pulp. It is dull grey in colour and relatively weak. It is available lined on one side with unbleached, semi or fully bleached chemical pulp. A range of such paperboards are available, with different quality liners. Chipboards are seldom used in direct contact with foods, but are used as outer cartons when the food is already contained in a film pouch or bag, e.g. breakfast cereals.

○ **Duplex board**: Duplex board is made from a mixture of chemical and mechanical pulp, usually lined on both sides with chemical pulp. It is used for some frozen foods, biscuits and similar products.

○ **Solid white board**: In solid white board, all plies are made from fully, bleached chemical pulp. It is used for some frozen foods, food liquids and other products requiring special protection.

2. Glass

Glass is 'an inorganic product of fusion which has cooled to a rigid state without crystallizing'. The atoms and molecules in glass have an amorphous random distribution. Scientifically this means that it has failed to crystallize from the molten state, and maintains a liquid-type structure at all temperatures. In appearance it is usually transparent but, by varying the components, this can be changed—as also can important properties such as thermal expansion, colour and the pH of aqueous extracts. Glass is hard and brittle, with a shell-like fracture.

Glass is primarily formed from oxides of metals, with the most common being dioxide which is common sand. Glass is made by mixing several naturally-occurring inorganic compounds at a temperature above their melting points.

The main ingredient is silica (sand) (SiO_2) that serves as the network-forming backbone of the glass. However, silica has a very high melting temperature, and molten silica has high viscosity that makes it difficult to form into shapes. Adding soda (Na_2O) modifies the silica network by disrupting some of the Si-O bonds, with resulting lower melting temperature and viscosity but reduced resistance to dissolving in water. Thus, lime (CaO) is added as a network stabilizer, with the result that durability is increased but tendency to crystallize is also increased. Finally, alumina (Al_2O_3) is added as an intermediate to resist crystallization. Minor amounts of colourants are added to produce coloured glass, including chromium oxide for green, cobalt oxide for blue, nickel oxide for violet, selenium for red, and iron plus sulfur and carbon for amber. Amber provides the best protection for light-sensitive foods and beverages, transmitting very little light with wavelength shorter than 450 nm.

Types of Glass

○ *White flint (clear glass)*: Colourless glass, known as white flint, is derived from soda, lime and silica. This composition also forms the basis for all other glass colours. A typical composition would be: Silica (SiO_2) 72%, from high purity sand; lime (CaO) 12%, from limestone (calcium carbonate); soda (Na_2O) 12%, from soda ash alumina (Al_2O_3), present in some of the other raw materials or in feldspar-type aluminous material; magnesia (MgO) and potash (K_2O), ingredients not normally added but present in the other materials. Cullet, recycled broken glass, when added to the batch reduces the use of these materials.

○ *Pale green (half white)*: Where slightly less pure materials are used, the iron content (Fe_2O_3) rises and a pale green glass is produced. Chromium oxide (Cr_2O_3) can be added to produce a slightly denser blue green colour.

○ *Dark green*: This colour is also obtained by the addition of chromium oxide and iron oxide.

○ *Amber*: Amber is usually obtained by melting a composition containing iron oxide under strongly reduced conditions. Carbon is also added. Amber glass has UV protection properties and could well be suited for use with light-sensitive products.

○ *Blue*: Blue glass is usually obtained by the addition of cobalt to a low-iron glass. Almost any coloured glass can be produced either by furnace operation or by glass colouring in the conditioning forehearth. The latter operation is an expensive way of producing glass and commands a premium product price. Forehearth colours would generally be outside the target price of most carbonated soft drinks.

Attributes/virtues of food packaged in glass containers

o *Quality image*: Consumer research by brand owners has consistently indicated that consumers attach a high quality perception to glass packaged products and they are prepared to pay a premium for them, for specific products such as spirits and liqueurs.

o *Transparency*: It is a distinct advantage for the purchaser to be able to see the product in many cases, e.g. processed fruit and vegetables.

o *Surface texture*: Most glass is produced with a smooth surface, other possibilities also exist, for example, for an overall roughened ice-like effect or specific surface designs on the surface, such as text or coats of arms. These effects emanate from the moulding but subsequent acid etch treatment is another option.

o *Colour*: A range of colours are possible based on choice of raw materials. Facilities exist for producing smaller quantities of nonmainstream colours.

o *Decorative possibilities*: Decorative possibilities including ceramic printing, powder coating, coloured and plain printed plastic sleeving and a range of labelling options.

o *Impermeability*: All practical purposes in connection with the packaging of food, glass is impermeable.

o *Chemical integrity*: Glass is chemically resistant to all food products, both liquid and solid. It is odourless.

o *Design potential*: Distinctive shapes are often used to enhance product and brand recognition.

o *Heat processable*: Glass is thermally stable, which makes it suitable for the hot-filling and the in-container heat sterilization and pasteurization of food products.

o *Microwaveable*: Glass is open to microwave penetration and food can be reheated in the container. Removal of the closures is recommended, as a safety measure, before heating commences, although the closure can be left loosely applied to prevent splashing in the microwave oven. Developments are in hand to ensure that the closure releases even when not initially slackened.

o *Tamper evident*: Glass is resistant to penetration by syringes. Container closures can be readily tamper-evidenced by the application of shrinkable plastic sleeves or in-built tamper evident bands. Glass can quite readily accept preformed metal and roll-on metal closures, which also provide enhanced tamper evidence.

o *Ease of opening*: The rigidity of the container offers improved ease of opening and reduces the risk of closure misalignment compared with plastic containers, although it is recognized that vacuum packed food products can be difficult to open. Technology in the development of lubricants in closure seals, improved application of glass surface treatments together with improved control of filling and retorting all combine to reduce the difficulty of closure removal. However, it is essential in order to maintain shelf life that sufficient closure torque is retained, to ensure vacuum retention with no closure back-off during processing and distribution.

o *UV protection*: Amber glass offers UV protection to the product and, in some cases, green glass can offer partial UV protection.

o *Strength*: Although glass is a brittle material glass containers have high top load strength making them easy to handle during filling and distribution. While the weight factor of glass is unfavourable compared with plastics, considerable savings are to be made in warehousing and distribution costs. Glass containers can withstand high top loading with minimal secondary packaging. Glass is an elastic material and will absorb energy.

3. Plastic

Plastic is an organic macromolecular compounds obtained by polymerisation, polycondensation, polyaddition or any similar process from molecules with a lower molecular weight or by chemical alteration of natural macromolecular compounds.

Plastics are used in the packaging of food because they offer a wide range of appearance and performance properties which are derived from the inherent features of the individual plastic material and how it is processed and used.

Plastics are resistant to many types of compound—they are not very reactive with inorganic chemicals, including acids, alkalis and organic solvents, thus making them suitable, i.e. inert, for food packaging.

Plastics do not support the growth of microorganisms.

Some plastics may absorb some food constituents, such as oils and fats, and hence it is important that a thorough testing is conducted to check all food applications for absorption and migration. Gases such as oxygen, carbon dioxide and nitrogen together with water vapour and organic solvents permeate through plastics. The rate of permeation depends on:

- o type of plastic
- o thickness and surface area
- o method of processing
- o concentration or partial pressure of the permeant molecule
- o storage temperature

Application of Plastic in Food Processing

Plastics are used as containers, container components and flexible packaging. In usage, by weight, they are the second most widely used type of packaging and first in terms of value.

Applications of plastic are:

- o Rigid plastic containers such as bottles, jars, pots, tubs and trays
- o Flexible plastic films in the form of bags, sachets, pouches and heat-sealable flexible lidding materials
- o Plastics combined with paperboard in liquid packaging cartons
- o Expanded or foamed plastic for uses where some form of insulation, rigidity and the ability to withstand compression is required
- o Plastic lids and caps and the wadding used in such closures
- o Diaphragms on plastic and glass jars to provide product protection and tamper evidence plastic bands to provide external tamper evidence

○ Pouring and dispensing devices to collate and group individual packs in multipacks, e.g. Hi-cone rings for cans of beer, trays for jars of sugar preserves, etc.

○ Plastic films used in cling, stretch and shrink wrapping

    ○ Films used as labels for bottles and jars, as flat glued labels or heat shrinkable sleeves

○ Components of coatings, adhesives and inks.

Types of Plastic used in Packaging

i. **Polyethylene (PE):** PE is structurally the simplest plastic and is made by addition polymerization of ethylene gas in a high temperature and pressure reactor. A range of low, medium and high density resins are produced, depending on the conditions (temperature, pressure and catalyst) of polymerization.

    Polyethylenes are readily heat sealable. They can be made into strong, tough films, with a good barrier to moisture and water vapour. They are not a particularly high barrier to oils and fats or gases such as carbon dioxide and oxygen compared with other plastics, although barrier properties increase with density. The heat resistance is lower than that of other plastics used in packaging, with a melting point of around 120°C, which increases as the density increases.

ii. **Polypropylene (PP):** PP is an addition polymer of propylene formed under heat and pressure using Zieger-Natta type catalysts to produce a linear polymer with protruding methyl (CH_2) groups. The resultant polymer is a harder and denser resin than PE and more transparent in its natural form.

    The high melting point of PP (160°C) makes it suitable for applications where thermal resistance is needed. The surfaces of PP films are smooth and have good melting characteristics. PP films are relatively stiff. When cast, the film is glass clear and heat sealable.

    It is used for presentation applications to enhance the appearance of the packed product. PP is chemically inert and resistant to most commonly found chemicals, both organic and inorganic. It is a barrier to water vapour and has oil and fat resistance.

iii. **Polyethylene terephthalate (PET):** PET can be made into film by blowing or casting. It can be blow moulded, injection moulded, foamed, extrusion coated on paperboard and extruded as sheet for thermoforming. PET can be made into a biaxially oriented range of clear polyester films produced on essentially the same type of extrusion and Stenter-orienting equipment as OPP.

PET melts at a much higher temperature than PP, typically 260°C, and due to the manufacturing conditions does not shrink below 180°C. This means that PET is ideal for high-temperature applications using steam sterilization, boiling-the-bag and for cooking or reheating in microwave or conventional radiant heat ovens. The film is also flexible in extremes of cold, down to −100°C. PET is a medium oxygen barrier on its own but becomes a high barrier to oxygen and water vapour when metalized with aluminium. This is used for vacuumised coffee and bag-in-box liquids, where it is laminated with EVA on both sides to produce highly effective seals. It is also used in snack food flexible packaging for products with high fat content requiring barriers to oxygen and ultra violet (UV) light.

PET film is also used as the outer reverse-printed ply in retort pouches, providing strength and puncture resistance, where it is laminated with aluminium foil and either PP or HDPE. PET can be oxide coated with SiO_2 to improve the barrier, whilst remaining transparent, retortable and microwaveable. PET is the fastest growing plastic for food packaging applications as a result of its use in all sizes of carbonated soft drinks and mineral water bottles which are produced by injection stretch blow moulding. PET bottles are also used for edible oils, as an alternative to PVC.

iv. **Ethylene vinyl acetate (EVA):** EVA is a copolymer of ethylene with vinyl acetate. It is similar to PE in many respects, and it is used, blended with PE, in several ways. The properties of the blend depend on the proportion of the vinyl acetate component. Generally, as the VA component increases, sealing temperature decreases and impact strength, low temperature flexibility, stress resistance and clarity increase. EVA is also a major component of hot melt adhesives, frequently used in packaging machinery to erect and close packs, e.g. folding cartons and corrugated packaging.

v. **Polyamide (PA):** Polyamides (PA) are commonly known as nylon. However, nylon is not a generic name; it is the brand name for a range of nylon products made by Dupont.

PA resins can be used to make blown film, and they can be coextruded. PA can be blended with PE, PET, EVA and EVOH. It can be blow moulded to make bottles and jars which are glass clear, low in weight and have a good resistance to impact. PA film is used in retortable packaging in structures such as PA/aluminium foil/PP. The film is non-whitening in retort processing. PA is relatively expensive compared with, for example, PE, but as it has superior properties, it is effective in low thicknesses.

vi. **Polyvinyl chloride (PVC):** PVC has excellent resistance to fat and oil. It is used in the form of blow moulded bottles for vegetable oil and fruit drinks. It has good clarity. As a film, it is tough, with high elongation, though with relatively low tensile and tear strength. The moisture vapour transmission rate is relatively high, though adequate for the packaging of mineral water, fruit juice and fruit drinks in bottles. PVC softens, depending on its composition, at relatively low temperatures (80–95°C).

It is plasticized, and the high stretch and cling make it suitable for overwrapping fresh produce, e.g. apples and meat in rigid trays using semi-automatic and manual methods.

Unplasticised PVC (UPVC) has useful properties but is a hard, brittle material, and modification is necessary for it to be used successfully. Flexibility can be achieved by the inclusion of plasticizers, reduced surface friction with slip agents, various colours by the addition of pigments and improved thermal processing by the addition of stabilizing agents.

vii. **Polystyrene (PS):** It is less well known as an oriented plastic film, though the film has interesting properties. It has high transparency (clarity). It is stiff, with a characteristic crinkle, suggesting freshness, and has a dead fold property. It has a low barrier to moisture vapour and common gases, making it suitable for packaging products, such as fresh produce, which need to breathe. PS is easily processed by foaming to produce a rigid lightweight material which has good impact protection and thermal insulation properties.

4. Metal

Two basic types of alloyed metals are used in food packaging, i.e. steel and aluminium.

Steel is used primarily to make rigid cans, whereas aluminium is used to make cans as well as thin aluminium foils and coatings.

Nearly all steel used for cans was coated with a thin layer of tin to inhibit corrosion, and called "tin can". The reason for using tin was to protect the metal can from corrosion by the food. Tin is not completely resistant to corrosion, but its rate of reaction with many food materials is considerably slower than that of steel.

The strength of the steel plate is another important consideration especially in larger cans that must withstand the pressure stresses of retorting, vacuum canning and other processes. Can strength is determined by the temper given the steel, the thickness of the plate, the size and the geometry of the can, and certain construction features such as horizontal ribbing to increase rigidity. This ribbing is known as beading. The user of cans will find it necessary to consult frequently with the manufacturer on specific applications, since metal containers like all other materials of packaging are undergoing constant change.

Aluminium is light weight, resistant to atmospheric corrosion, and can be shaped or formed easily. However, aluminium has considerably less structural strength than steel at the same gauge thickness. This means that aluminium has limited use in cans such as those used with retorted foods. Aluminium works well in very thin beverages cans that contain internal pressure such as soda or beer. This internal pressure from CO_2 gives rigidity to the can. Aluminium in contact with air forms an aluminium oxide film which is resistant to atmospheric corrosion. However, if the oxygen concentration is low, as it is within most foods containing cans, this aluminium oxide film gradually becomes depleted and the underlying aluminium metal is then no longer highly resistant to corrosion (potter).

Metals used in Packaging

The metal materials used in food packaging are aluminium, tinplate and electrolytic chromium-coated steel (ECCS). Aluminium is used in the form of foil or rigid metal.

o **Aluminium foil**: Aluminium foil is produced from aluminium ingots by a series of rolling operations down to a thickness in the range of 0.15–0.008 mm. Most foil used in packaging contains not less than 99.0% aluminium, with traces of silicon, iron, copper and in some cases, chromium and zinc. Foil used in semi-rigid containers also contains up to 1.5% manganese. After rolling, foil is annealed in an oven to control its ductility. This enables foils of different tempers to be produced from fully annealed (dead folding) to hard, rigid material. Foil is a bright, attractive material, tasteless, odorless and inert with respect to most food materials. For contact with acid or salty products, it is coated with nitrocellulose or some polymer material. It is mechanically weak, easily punctured, torn or abraded. Foil is used as a component in laminates, together with polymer materials and, in some cases, paper. These laminates are formed into sachets or pillow packs on FFS equipment. Examples of foods packaged in this way include dried soups, sauce mixes, salad dressings and jams. Foil is included in laminates used for retort pouches and rigid plastic containers for ready meals. It is also a component in cartons for UHT milk and fruit juices.

o **Tin**

• *Tinplate:* Tinplate is the most common metal material used for food cans. It consists of a low-carbon, mild steel sheet or strip, 0.50–0.15 mm thick, coated on both sides with a layer of tin. This coating seldom exceeds 1% of the total thickness of the tinplate. The mechanical strength and fabrication characteristics of tinplate depend on the type of steel and its thickness. The minor constituents of steel are carbon, manganese, phosphorous, silicon, sulfur and copper. At least four types of steel, with different levels of these constituents, are used for food cans. The corrosion resistance and appearance of tinplate depend on the tin coating.

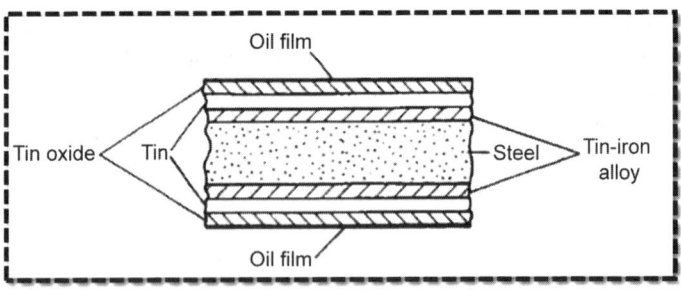

Fig. 12.1: Structure of tin plate

- *Tin coating*: The role of tin coating is an essential component of the can construction and plays an active role in determining shelf life. The most significant aspect of the role of the tin coating is that it protects the steel base-plate which is the structural component of the can.

  Without a coating of tin, the exposed iron would be attacked by the product and this would cause serious discolouration and off-flavours in the product and swelling of the cans; in extreme cases the iron could be perforated and the cans would lose their integrity.

  The second role of tin is that it provides a chemically reducing environment, any oxygen in the can at the time of sealing being rapidly consumed by the dissolution of tin. This minimizes product oxidation and prevents colour loss and flavour loss in certain products.

- *Tin toxicity:* High concentrations of tin in food irritate the gastrointestinal tract and may cause stomach upsets in some individuals, with symptoms which include nausea, vomiting, diarrhoea, abdominal cramps, abdominal bloating, fever and headache. Tin corrosion occurs throughout the shelf life of the product. It is therefore imperative to take steps to reduce the rate of corrosion. Accelerating factors include heat, oxygen, nitrate, some chemical preservatives and dyes, and certain particularly aggressive food types (e.g. celery, rhubarb). A high vacuum level is one effective method of reducing the rate of tin pick-up in cans with un-lacquered components.

o **Electrolytic Chromium-Coated Steel (ECCS):** Electrolytic chromium-coated steel (ECCS), sometimes described as tin-free steel, is finding increasing use for food cans. It consists of low-carbon, mild CR or DR steel coated on both sides with a layer of metallic chromium and chromium oxide, applied electrolytically. ECCS is less resistant to corrosion than tinplate and is normally lacquered on both sides. It is more resistant to weak acids and sulfur staining than tinplate.

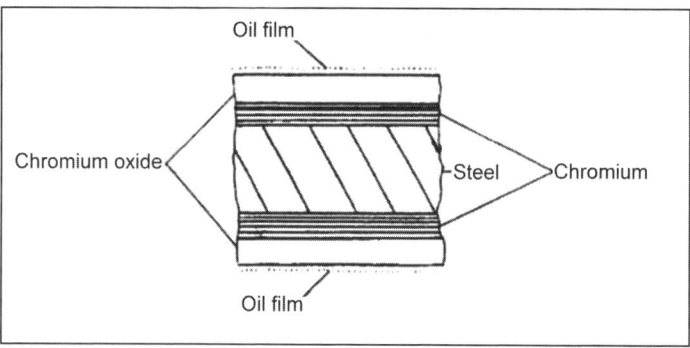

Fig. 12.2: Structure of ECCS plate

o **Aluminium alloy:** Hard-temper aluminium alloy, containing 1.5–5.0% magnesium, is used in food can manufacture. It is lighter but mechanically weaker than tinplate. It is manufactured in a similar manner to aluminium foil. It is less resistant to corrosion than tinplate and needs to be lacquered for most applications.

o **Lead:** Lead was a problem with older, soldered cans but levels are now very low. However, some tinplate is contaminated with minimal amounts of lead. The manufacture of lead soldered cans may still be found in the developing world.

o **Lacquers:** The presence of lacquer or enamel very effectively limits dissolution of tin into the product, and so the use of lacquers is becoming increasingly common, even with those products which were previously packed in plain tinplate cans. There are several different types of lacquer in common use today. By far the most common type is the Epoxy phenolic group, which are suitable for packing meat, fish, vegetables and fruit products. These have largely replaced the oleoresinous group, which had a similar wide range of application.

Some canners use cans lacquered with vinyl resins, which have the important quality of being free from any taste and odour, and are therefore particularly suitable for dry packs such as biscuits and powders, but also some drinks. White vinyl lacquers have been used where staining of the underlying metal caused by reaction with the product is a problem. Also, white vinyl lacquers have been used for marketing reasons in order to present a hygienic/clinical appearance and not the aesthetically undesirable corrosion patterns on tinplate.

5. Edible Films

Edible films and coatings formed from polysaccharides, proteins, lipids, resins, and/or waxes fall within the active packaging definition, since they can enhance the protective function, provide convenience, and minimize package environmental impact.

Edible films placed or formed between components of a packaged food control transfer of moisture, oils, etc. over which the package has no control. Edible coatings or edible film pouches (as a primary package) work to complement the protective function of the non-edible (secondary) package. Such coatings and films can act as barriers to the external environment and maintain food integrity, thus reducing the amount of packaging required.

Edible film pouches carrying premeasured amounts of ingredients can provide the convenience of placing pouch with ingredients into the food formulation. Edible coatings can also carry antimicrobials that can inhibit microbial growth at both the food-coating interface and the coating outer surface.

Food materials can be protected from loss of volatiles or reaction with other food ingredients by being encapsulated in protective edible materials. This can be done by spray drying various flavouring materials emulsified with gelatin, gum Arabic, or other edible materials to form a thin protective coating around each food particle. The coatings of raisins with starches to prevent them from moistening a packaged breakfast cereal and the coating of nuts with mono-glyceride derivatives to protect them from oxidative rancidity are additional examples of edible coatings.

Edible films are also used to coat fresh fruits and vegetables to reduce moisture loss and to provide increased resistance to growth of surface molds.

The most common and oldest edible film is **wax**. A wide range of products such as apples are waxed for appearance and improved keeping quality. Newer edible films are being developed which can keep produce longer.

6. Retortable Pouches and Trays

Flexible pouches, semi-rigid/rigid plastic trays and cans, and paperboard-based cartons have been developed as alternatives to heat processing (retorting) in rigid metal cans or glass containers. The pouches, trays, and tubs are always multilayer laminate structures that contain different polymers which provide heat resistance, strength, and toughness (PET), pierce and pinhole resistance (nylon), oxygen barrier (EVOH, nylon or PVDC) and (for the pouches and trays) heat sealability (PP). An aluminium foil layer often serves as the moisture and oxygen barrier in pouches. The retortable paperboard cartons have external and internal PP layers that are impermeable to liquid and allow heat sealing, along with an internal aluminium layer that provides a gas and light barrier.

Retortable pouches can be either preformed or in-line formed using form/fill/seal equipment. Common pouch structures are PET/nylon/foil/PP and PET/nylon/EVOH or PVDC/PP. Retortable trays have a semi-rigid or rigid body and a sealable flexible lid. The trays are generally made from coextruded laminate such as PET/EVOH/PP by thermoforming. Retortable tubs are made from similar multi-layer laminates. An easy-open scored metal lid with pull ring is double seamed onto the tub body.

The advantage of retortable pouches and trays is that they have thinner profile than conventional metal or glass containers. The results are shortened process times, reduced energy consumption, and improved food quality due to more rapid and even heat transfer. In addition, retort pouches, trays, and tubs are convenient because of easy transport (due to shape and light weight) and easy opening. Plastic (with no foil layer) pouches, trays, and tubs are microwaveable. The main disadvantage of retortable pouches, trays, tubs, and cartons is more difficult recycling.

Flexible materials can be combined to withstand even the adverse conditions of retorting encountered with low-acid foods. Such "flexible cans" have becomes standard containers for some applications such as providing foods to soldiers in the fields. The advantages of pouches and trays over cans and jars of equivalent volume include shorter retort times, which can produce higher quality products and save on energy, lighter weight, increased compactness, easier opening and easier disposability. Retortable pouches are constructed of a three-ply laminate consisting of 1. an outer layer of polyester films for high-temperature resistance, strength and printability. 2. A middle layer of aluminium foil for barrier properties, and 3. an inner layer of polypropylene film that provides heat- seal integrity. Retortable trays are constructed from multilayers of polymers, one of which is ethylene- vinyl alcohol to provide an oxygen barrier. These trays are often sealed with a polymer-foil laminate film.

7. Cloth Materials

Jute and cotton are woven materials which have been used for packaging foods. Sacks made of jute are used, to a limited extent, for fresh fruit and vegetables, grains and dried legumes. However, multiwall paper sacks and plastic sacks have largely replaced them for such products. Cotton bags have been used in the past for flour, sugar, salt and similar products. Again, paper and plastic bags are now mainly used for these foods.

Cotton scrims are used to pack fresh meat. However, synthetic materials are increasingly used for this purpose.

8. Wooden Containers

Outer wooden containers are used when a high degree of mechanical protection is required during storage and transport. They take the form of crates and cases. Wooden drums and barrels are used for liquid products. The role of crates has largely been replaced by shipping containers. Open cases find limited use for fish, fruits and vegetables, although plastic cases are now widely used. Casks, kegs and barrels are used for storage of wines and spirits. Oak casks are used for high quality wines and spirits. Lower quality wines and spirits are stored in chestnut casks.

9. Composite Containers

So called composite containers usually consist of cylindrical bodies made of paperboard or fiberboard with metal or plastic ends. Where good barrier properties are required, coated or laminated board may be used for the body or aluminium foil may be incorporated into it. Small containers, less than 200 mm in diameter, are referred to as tubes or cans and are used for foods such as salt, pepper, spices, custard powders, chocolate beverages and frozen fruit juices. Larger containers, known as fiberboard drums, are used as alternatives to paper or plastic sacks or metal drums for products such as milk powder, emulsifying agents and cooking fats.

10. Regenerated Cellulose

Regenerated cellulose (cellophane) differs from the polymer films in that it is made from wood pulp. Good quality, bleached sulphite pulp is treated with sodium hydroxide and carbon disulphide to produce sodium cellulose xanthate. This is dispersed in sodium hydroxide to produce viscose. The viscose is passed through an acid-salt bath which salts out the viscose and neutralizes the alkali. It provides general protect against dust and dirt, some mechanical protection and is greaseproof. When dry it is a good barrier to gases, but becomes highly permeable when wet. Plain cellulose is little used in food packaging. Plain regenerated cellulose is mainly used coated with various materials which improve its functional properties. The most common coating material is referred to as 'nitrocellulose' but is actually a mixture of nitrocellulose, waxes, resins, plasticizers and some other agents.

11. Cellulose Acetate

Cellulose acetate is made from waste cotton fibers which are acetylated and partially hydrolyzed. The film is made by casting from a solvent or extrusion. It is clear, transparent and has a sparkling appearance. It is highly permeable to water vapour, gases and volatiles. It is not much used in food packaging except as window material in cartons. It can be thermoformed into semirigid containers or as blister packaging.

SELECTION CRITERIA OF PACKAGING MATERIAL FOR THE RAW AND PROCESSED FOODS

Mechanical Damage

Fresh, processed and manufactured foods are susceptible to mechanical damage. The bruising of soft fruits, the break-up of heat processed vegetables and the cracking of biscuits are examples. Such damage may result from sudden impacts or shocks during handling and transport, vibration during transport by road, rail and air and compression loads imposed when packages are stacked in warehouses or large transport vehicles.

Appropriate packaging can reduce the incidence and extent of such mechanical damage. Packaging alone is not the whole answer. Good handling and transport procedures and equipment are also necessary.

The selection of a packaging material of sufficient strength and rigidity can reduce damage due to compression loads.

Metal, glass and rigid plastic materials may be used for primary or consumer packages. Fiberboard and timber materials are used for secondary or outer packages. The incorporation of cushioning materials into the packaging can protect against impacts, shock and vibration. Corrugated papers and boards, pulp board and foamed plastics are examples of such cushioning materials. Restricting movement of the product within the package may also reduce damage. This may be achieved by tight-wrapping or shrink-wrapping. Inserts in boxes or cases or thermoformed trays may be used to provide compartments for individual items such as eggs and fruits.

Permeability Characteristics

The rate of permeation of water vapour, gases (O_2, CO_2, N_2, and ethylene) and volatile odour compounds into or out of the package is an important consideration, in the case of packaging films, laminates and coated papers.

Foods with relatively high moisture contents tend to lose water to the atmosphere. This results in a loss of weight and deterioration in appearance and texture. Meat and cheese are typical examples of such foods. Products with relatively low moisture contents will tend to pick up moisture, particularly when exposed to a high humidity atmosphere. Dry powders such as cake mixes and custard powders may cake and lose their free flowing characteristics. Biscuits and snack foods may lose their crispness. If the water activity of a dehydrated product is allowed to rise above a certain critical level, microbiological spoilage may occur.

In such cases a packaging material with a low permeability to water vapour, effectively sealed, is required. In contrast, fresh fruit and vegetables continue to respire after harvesting. They use up oxygen and produce water vapour, carbon dioxide and ethylene. As a result, the humidity inside the package increases. If a high humidity develops, condensation may occur within the package when the temperature fluctuates. In such cases, it is necessary to allow for the passage of water vapour out of the package. A packaging material which is semipermeable to water vapour is required in this case.

The shelf life of many foods may be extended by creating an atmosphere in the package which is low in oxygen. This can be achieved by vacuum packaging or by replacing the air in the package with carbon dioxide and/or nitrogen. Cheese, cooked and cured meat products, dried meats, egg and coffee powders are examples of such foods. In such cases, the packaging material should have a low permeability to gases and be effectively sealed. If a respiring food is sealed in a gastight container, the oxygen will be used up and replaced with carbon dioxide. The rate at which this occurs depends on the rate of respiration of the food, the amount in the package and the temperature.Over a period of time, an anaerobic atmosphere will develop inside the container. If the oxygen content falls below 2%, anaerobic respiration will set in and the food will spoil rapidly. The influence of the level of carbon dioxide in the package varies from product to product. Some fruits and vegetables can tolerate, and may even benefit from, high levels of carbon dioxide while others do not. In such cases, it is necessary to select a packaging material which permits the movement of oxygen into and carbon dioxide out of the package, at a rate which is optimum for the contents. Ethylene is produced by respiring fruits. Even when present in low concentrations, this can accelerate the ripening of the fruit. The packaging material must have an adequate permeability to ethylene to avoid this problem.

To retain the pleasant odor associated with many foods, such as coffee, it is necessary to select a packaging material that is a good barrier to the volatile compounds which contribute to that odor. Such materials may also prevent the contents from developing taints due to the absorption of foreign odors. It is worth noting here that films that are good barriers to water vapour may be permeable to volatiles. In those cases where the movement of gases and vapours is to be minimized, metal and glass containers, suitably sealed, may be used. Many flexible film materials, particularly if used in laminates, are also good barriers to vapours and gases. Where some movement of vapours and/or gases is desirable, films that are semi-permeable to them may be used. For products with high respiration rates the packaging material may be perforated.

Greaseproof

In the case of fatty foods, it is necessary to prevent egress of grease or oil to the outside of the package, where it would spoil its appearance and possibly interfere with the printing and decoration. Greaseproof and parchment papers may give adequate protection to dry fatty foods, such as chocolate and milk powder, while hydrophilic films or laminates are used with wet foods, such as meat or fish.

Temperature

A package must be able to withstand the changes in temperature which it is likely to encounter, without any reduction in performance or undesirable change in appearance. This is of particular importance when foods are heated or cooled in the package.

For many decades, metal and glass containers were used for foods which were reheated in the package. It is only in relatively recent times that heat resistant laminates were developed for this purpose. Some packaging films become brittle when exposed to low temperatures and are not suitable for packaging frozen foods. The rate of change of temperature may be important. For example, glass containers have to be heated and cooled slowly to avoid breakage.

Light

Many food components are sensitive to light, particularly at the blue and ultraviolet end of the spectrum. Vitamins may be destroyed, colours may fade and fats may develop rancidity when exposed to such light waves. The use of packaging materials which are opaque to light will prevent these changes. If it is desirable that the contents be visible, for example, to check the clarity of a liquid, coloured materials which filter out short wavelength light may be used. Amber glass bottles, commonly used for beer in the UK, perform this function. Pigmented plastic bottles are used for some health drinks.

CHEMICAL COMPATIBILITY OF THE PACKAGING MATERIAL AND THE CONTENTS OF THE PACKAGE

It is essential in food packaging that no health hazard to the consumer should arise as a result of toxic substances, present in the packaging material, leaching into the contents. In the case of flexible packaging films, such substances may be residual monomers from the polymerization process or additives such as stabilizers, plasticizers, colouring materials, etc.

To establish the safety of such packaging materials two questions need to be answered:

a. Are there any toxic substances present in the packaging material/

b. Will they leach into the product?

Toxicological testing of just one chemical compound is lengthy, complicated and expensive, usually involving extensive animal feeding trials and requiring expert interpretation of the results.

Protection against Microbial Contamination

Another role of the package may be to prevent or limit the contamination of the contents by microorganisms from sources outside the package. This is most important in the case of foods that are heat-sterilized in the package, where it is essential that post-process contamination does not occur. The metal can has dominated this field for decades and still does. The reliability of the double seam in preventing contamination is one reason for this dominance. Some closures for glass containers are also effective barriers to contamination. It is only in relatively recent times that plastic containers have been developed, which not only withstand the rigors of heat processing, but also whose heat seals are effective in preventing post-process contamination. Effective seals are also necessary on cartons, cups and other containers which are aseptically filled with UHT products. The sealing requirements for containers for pasteurized products and foods preserved by drying, freezing, curing, etc. are not so rigorous. However, they should still provide a high level of protection against microbial contamination.

In-Package Microflora

The permeability of the packaging material to gases and the packaging procedure employed can influence the type of microorganisms that grow within the package. Packaging foods in materials that are highly permeable to gases is not likely to bring about any significant change in the microflora, compared to unpackaged foods. However, when a fresh or mildly processed food is packaged in a material that has a low permeability to gases and when an anaerobic atmosphere is created within the package, as a result of respiration of the product or because of vacuum or gas packaging, the type of micro-organisms that grow inside the package are likely to be different to those that would grow in the unpackaged food. There is a danger that pathogenic microorganisms could flourish under these conditions and result in food poisoning. Such packaging procedures should not be used without a detailed study of the microbiological implications, taking into account the type of food, the treatment it receives before packaging, the hygienic conditions under which it is packaged and the temperature at which the packaged product is to be stored, transported, displayed in the retail outlet and kept in the home of the consumer.

SELECTION CRITERIA OF PACKAGING MATERIAL FOR RAW FOODS

Raw Meat

The adoption of preservative packaging for raw meats has led to major changes in the processing and marketing of such products.

As a result of the widespread adoption of vacuum packaging for primal cuts of red meats, trade in red meat carcasses has declined to trivial proportions in many developed countries, and the international trade in chilled raw meats has greatly increased, with a consequent decline in trading of frozen meats.

The enhanced stability of vacuum-packaged products has facilitated consolidation of meat-packing facilities. Most meat is offered to consumers in a freshly or recently cut form, with little further processing to suppress the normal microbiological flora present from the contamination received during the killing and breaking operations required to reduce carcass meat to edible cuts. Fresh meat is vulnerable to microbiological deterioration from microorganisms. These microorganisms can be as benign as slime formers to stink producers to pathogens such as *E. coli* O157:H7. The major mechanisms to retard fresh meat spoilage are temperature reduction, often coupled with reduced oxygen during distribution, to retard normal spoilage microbial growth. Reduced oxygen also leads to fresh meat colour being the purple of myoglobin, a condition changed upon exposure to air which converts the natural meat pigment to bright cherry red oxy-myoglobin characteristic of most fresh meat offered to and accepted by consumers.

Fish and Sea Foods

Varieties of fish are among the most difficult of all foods to preserve in their fresh state because of their inherent microbiological populations many of which are psychrophilic, i.e. capable of growth at refrigerated temperatures. Further, seafood may harbor a nonproteolytic anaerobic pathogen, *Clostridium botulinum* type E, capable of toxin production without signaling spoilage.

The high-quality shelf life of most seafood in chill storage is relatively short, being only a few days. This short period does not allow sufficient time from reception through to distribution and display to ensure the restaurateur or consumer can obtain seafood at its best.

Packaging for fresh seafood is generally moisture resistant but not necessarily against microbial contamination. Simple polyethylene film is employed often as a liner in corrugated fiberboard cases. The polyethylene serves not only to retain product moisture but also to protect the structural case against internal moisture.

Seafood may be frozen in which case the packaging is usually a form of moisture resistant material plus structure such as polyethylene pouches or polyethylene coated paperboard cartons.

Canning of seafood is much like that for meats because all seafoods are low acid and so require high pressure cooking or retorting to effect sterility in hermetically sealed metal cans.

Fruits and Vegetables

Increasing demand for a wide range of harvested fruits or vegetables (raw and fresh-cut) has led to dynamic growth in sales and new market opportunities for the fresh produce sector. However, their preservation still constitutes one of the most challenging applications for the food industry.

Fresh produce is a living, "breathing" entity fostering the physiological consumption of oxygen and production of carbon dioxide and water vapour. From a spoilage standpoint, fresh produce is more subject to physiological than to microbiological spoilage, and measures to extend the shelf life are designed to retard such reactions and water loss.

One major problem is that produce may enter into respiratory anaerobiosis if the oxygen concentration is reduced to near extinction. In respiratory anaerobiosis, the pathways produce undesirable flavour compounds.

To minimize the production of these undesirable end products, elaborate packaging systems have been and continue to be developed. Most of these involve mechanisms to permit air into the package to compensate for the oxygen consumed by the respiring produce. High gas permeability plastic films, micro perforated plastic films, plastic films disrupted with mineral fill, and films fabricated from temperature-sensitive polymers have all been proposed or used commercially.

Fresh-cut vegetables, especially lettuce, cabbage, and carrots have been a major product in both the retail and the hotel/restaurant/institutional market. Cleaning, trimming and size reduction lead to greater surface to volume of the produce and to the expression of fluids from the interior to increase the respiration and microbiological growth rate. On the other hand, commercial fresh-cutting operations generally are far superior to mainstream fresh produce handling in cleanliness, speed through the operations, temperature reduction, and judicious application of microbicides such as chlorine. Uncut produce packaging is really a multitude of materials, structures, and forms that range from the old and traditional, such as wood crates, to inexpensive, such as injection-molded polypropylene baskets, to polyethylene liners within waxed corrugated fiberboard cases. Much of the packaging is designed to help retard moisture loss from the fresh produce or to resist the moisture evapourating or dripping from the produce (or, occasionally, its associated ice) to ensure the maintenance of the structure throughout distribution. Some packaging recognizes the issue of anaerobic respiration and incorporates deliberate openings to ensure passage of air into the package, for example, perforated polyethylene pouches for apples or potatoes.

For freezing, vegetables are cleaned, trimmed, cut, and blanched prior to freezing and then packaging, or prior to packaging and then freezing. Blanching and the other processing operations reduce the number of microorganisms. Produce may be individually quick frozen (IQF) using cold air or cryogenic liquids prior to packaging or frozen after packaging as in folding paperboard cartons. Frozen food packages are generally relatively simple monolayer polyethylene pouches or polyethylene-coated paperboard to retard moisture loss. Fresh-cut vegetables coupled with bread sticks and dip constitutes a reasonably flavourful mouthful and nutritious snack for adults and younger persons.

Milk

Milk is a complex mixture of water, proteins, lipids, carbohydrates, enzymes, vitamins, and minerals. Due to its specific composition and a pH close to neutral, it is a highly perishable product with high spoilage potential that can result in rapid deterioration of quality and safety. Packaging serves a number of different functions, including containment, protection, convenience, and communication, the most important being protection. Packaging protects milk and dairy products against environmental, physical, chemical, as well as mechanical hazards. It also protects the product from loss of desirable flavour compounds or pick-up of undesirable odours, and contamination from spoilage or pathogenic microorganisms, insects, or rodents during storage and distribution. An effective packaging system should fulfill numerous other requirements, including compatibility with the dairy product it contains recyclability or reuse, tamper evidence, nontoxicity, aesthetics, machinability, and functionality in terms of shape, size, and disposability.

Milk and its derivatives are generally excellent microbiological growth substrates and therefore potential sources for pathogens. For this reason, almost all milk is thermally pasteurized or heated short of sterility as an integral element of processing. Refrigerated distribution is generally dictated for all products that are pasteurized, to minimize the probability of spoilage. In recent years, milk packaging has been upgraded to incorporate reclosure, a feature that has been missing from gable top polyethylene-coated paperboard cartons. Further, in recent years, the packaging environmental conditions have been upgraded microbiologically to enhance refrigerated shelf life. Aseptic packaging is employed to deliver ambient temperature shelf stable fluid dairy products. The most common processing technology is ultra high temperature short time thermal treatment to sterilize the product, followed by aseptic transfer into the packaging equipment. Fluid milk is generally pasteurized, cooled, and filled into bag-in-box pouches for refrigerated distribution.

PACKAGING GRAPHICS AND LABELLING

Printing

Printing is a process in which text and images are reproduced, typically with ink on paper using a printing press made from letters, photographs and drawing. It is often carried out as a large-scale industrial process, and is an essential part of publishing and transaction printing. The basic systems of printing are: (1) the original, (2) the plate, (3) printing ink, (4) a printing medium such as paper and (5) a printing machine. Printing can be classified into two parts:

1. **Direct printing:** In this type of printing, printed material comes in direct contact with the plate so that the ink is directly applied to the printing medium.
2. **Indirect printing:** In this printing, the ink is first applied to the blanket cylinder from the plate and then printing medium comes in contact with the blanket.

Printing Technologies

Numbers of printing are in trend for printing purposes of packaging materials which are as follows:

1. **Lithography**: Lithography is a method in which printing is applied on a smooth surface. Lithography is a printing process that uses chemical processes to create an image.

Lithography is a form of planographic printing, meaning that the surface is flat, in contrast to relief printing or intaglio printing. For instance, the positive part of an image would be a hydrophobic chemical, while the negative image would be water. Thus, when the plate come in contact with a compatible ink and water mixture, the ink will adhere to the positive image and the water will clean the negative image. This allows for a relatively flat print plate which allows for much longer runs than the older physical methods of imaging (e.g. embossing or engraving). High-volume lithography is used today to produce packaging materials, just about any smooth, mass-produced item with print and graphics on it. Most books, indeed all types of high-volume text, are now printed using offset lithography.

2. **Colour printing:** Chromolithography is a method for making multi-colour prints. This type of colour printing stemmed from the process of lithography, and it includes all types of lithography that are printed in colour. Lithographers sought to find a way to print on flat surfaces with the use of chemicals instead of relief or intaglio printing.

   Chromolithographs are mainly used today as fine art instead of advertisements, and they are hard to find owing to poor methods of preservation and also because a cheaper form of printing replaced it. Many chromolithographs have deteriorated because of the acidic frames surrounding them. As stated earlier, production costs of chromolithographs were low, but efforts were still being made to find a cheaper way to mass produce coloured prints. Although purchasing a chromolithograph may have been cheaper than purchasing a painting, it was still expensive in comparison to other colour printing methods which were later developed.

3. **Screen printing:** Screen printing has its origins in simple stenciling, most notably of the Japanese form (katazome), used who cut banana leaves and inserted ink through the design holes on textiles, mostly for clothing. This was taken up in France.

4. **Flexography:** Flexographic printing is widely used in western countries. Flexography (also called "surface printing"), often abbreviated to "flexo", is a method of printing most commonly used for packaging (labels, tape, bags, boxes, banners, and so on). A flexo print is achieved by creating a mirrored master of the required image as a 3D relief in a rubber or polymer material. A measured amount of ink is deposited upon the surface of the printing plate. The print surface then rotates, contacting the print material which transfers the ink. Ink is picked up by a cavitated anilox roll and transferred to the printing plate. The ink is then transferred to the film. Because the costs of producing the plates are relatively low, flexographic printing is cost effective, especially for short runs. Its printing quality is inferior to that of modern printing techniques such as offset printing and gravure printing. Nowadays this printing technology is utilized for printing on polyethylene bags, corrugated boxes and carton after using the photosensitive resins of improved printing quality.

5. **Digital press:** Digital printing is the reproduction of digital images on a physical surface, such as common or photographic paper or paperboard-cover stock, film, cloth, plastic, vinyl, magnets, labels etc. It is now possible to create artwork on a computer and transfer the image directly to the packaging film. A design is created on a computer; it may be an individual design or replicated to give several hundreds of impressions. The ink, usually in powder form, is attracted on to the film surface and cured in place. Special coatings are necessary to receive the ink. A standard heat-sealable coating on the reverse side allows the film to be made immediately into packages.

It can be differentiated from litho, flexography, gravure or letterpress printing in many ways, some of which are;

i. Every impression made onto the paper can be different, as opposed to making several hundred or thousand impressions of the same image from one set of printing plates, as in traditional methods

ii. The ink or toner does not absorb into the substrate, as does conventional ink, but forms a layer on the surface and may be fused to the substrate by using an inline fuser fluid with heat process (toner) or UV curing process (ink).

iii. It generally requires less waste in terms of chemicals used and paper wasted in set up or make ready.

iv. It is excellent for rapid prototyping, or small print runs which means that it is more accessible to a wider range of designers and more cost effective.

6. **Frescography:** Frescography is a method for reproduction/creation of murals using digital printing methods. The frescography is based on digitally cut-out motifs which are stored in a database. CAM software programs then allow entering the measurements of a wall or ceiling to create a mural design with low resolution motifs. Since architectural elements such as beams, windows or doors can be integrated, the design will result in an accurately and tailor-fit wall mural. Once a design is finished, the low resolution motifs are converted into the original high resolution images and are printed on canvas by Wide-format printers. The canvas then can be applied to the wall in a wall-paperhanging like procedure and will then look like on-site created mural.

7. **3D printing:** Three-dimensional printing is a method of converting a virtual 3D model into a physical object. 3D printing is a category of rapid prototyping technology. 3D printers typically work by 'printing' successive layers on top of the previous to build up a three-dimensional object. 3D printers are generally faster, more affordable and easier to use than other additive fabrication technologies.

Modern Printing Technologies

1. **Offset printing:** Offset printing is widely used for printing on folding cartons in the packaging field. Offset printing is a widely used printing technique where the inked image is transferred (or "offset") from a plate to a rubber blanket, then to the printing surface. When used in combination with the lithographic process, which is based on the repulsion of oil and water, the offset technique employs a

flat image carrier on which the image to be printed obtains ink from ink rollers, while the non-printing area attracts a film of water, keeping the non-printing areas ink-free.

Offset printing has the following advantages:

1. Plate making time is short in this printing technology.
2. Plate making cost is less than that of gravure printing.
3. It is suitable for multicolour printing.
4. In offset printing large-sized plates are easily made.

2. **Gravure:** Gravure printing gives good quality of printed matter, either in monocolour or in four colours. Gravure printing is an intaglio printing technique, where the image to be printed is made up of small depressions in the surface of the printing plate. The cells are filled with ink and the excess is scraped off the surface with a doctor blade, then a rubber-covered roller presses paper onto the surface of the plate and into contact with the ink in the cells. The printing plates are usually made from copper and may be produced by digital engraving or laser etching. Gravure printing also has a high printing speed and is suitable for high volume printing. In packaging, gravure printing is performed on most flexible packages using cellophane, plastic film or aluminium foil. Gravure printing is used for long, high-quality print runs such as magazines, mail-order catalogues, packaging, and printing onto fabric and wallpaper. It is also used for printing postage stamps and decorative plastic laminates, such as kitchen worktops.

LABELLING AND LAMINATION

Label is the first point of contact between the consumer and the producer. It allows the consumers to know what exactly they are buying in terms of calories, proteins, fats, etc. and thus enables them to make a 'health conscious selection'. It informs the consumers regarding weight of the product, best before date, storage conditions and cooking recipe, if any. It allows consumers to compare food products by Value for Money. A label is a piece of paper, polymer, cloth, metal, or other material affixed to a container or article, on which is printed a legend, information concerning the product, addresses, etc. A label may also be printed directly on the container or article. Labels have many uses: Product identification, name tags, advertising, warnings, and other communication. Special types of labels called digital labels (printed through a digital printing) can also have special constructions such as RFID tags, security printing, and sandwich process labels.

Purpose of Labels

Information about Packaged Foods: It requires that all packaged foods list the name and address of the food's manufacturer, the weight or count of the food and nutrition facts for the food. The NLEA applies to all foods except for meat, poultry, eggs, prepared food or foods that are sold in bulk.

Nutrition value of product: The Nutrition Facts Label is the label with the most information for consumers. The first line of this label lists the serving size. The nutritional information that follows is based on this specific serving size. The next line lists the total calories, and the amount of calories that are from fat. The following lines contain the food's total fat content (including a breakout of saturated and trans fats), cholesterol and sodium. Carbohydrates, fiber, sugars, vitamins and minerals are listed next. The percent of the daily value for each nutrient, based on a 2,000 calorie diet, is listed on the right side of the label. The footnote on the bottom of the label has the FDA's recommended dietary guidelines. If the food label is very small, this footnote is abbreviated.

Decoration: When food product is choicely labeled in bright and attractive colours, it attracts consumers to buy. It acts as a silent shelf salesman. The colour and design should be in symmetry with product colour and the level should have some relationship to the size and shape of the package and container.

Warning: Food labels also having warning and instructions about the food product. Labels educate consumers about allergens, preparation methods and storage conditions for products.

Identification: Identification of the product is the main role of the labels as the consumer must be able to identify.

Name and address of the manufacturer, packer and/or seller and brand name also identify by the labels.

Types of Labels

Paper labels may be classified into four main categories:

1. Plain paper labels
2. Pre-gummed paper labels
3. Thermoplastic labels
4. Pressure sensitive paper labels

Plain paper labels: Plain paper labels are cheaper than other same quality of labels. Any type of the plain paper label can be printed by standard printing machine and by normal printing methods. These labels can be applied by simple hand application, semi-automatic to fully automatic procedure.

Pre-gummed paper labels: Pre-gummed paper labels are prepared from dextrin and gum Arabic coated papers and then calendared, flattened or non-curled by a special process. The advantage over plain paper is that they require only to be moistened with water for ready use as a postage stamp.

Thermoplastic labels: Thermoplastic labels are prepared from paper coated with a synthetic resin which melts and becomes tacky on the application of heat. There are two varieties (a) instant tack and (b) delayed action. In thermoplastic labels printing inks are used for the printing purpose. In these types of levels liquid phase or solvent does not activate the adhesive and should be heat resistant.

Table 12.2: Some of the common terms for labels

Back label	Used on back of containers.
Band label	Wraps around container or product, does not cover the entire surface.
Can label	Used on tin cylindrical containers.
Die cut label	Label of irregular shape cut with a die.
Embossed label	Labels which have three-dimensional effects
End label	Essentially a spot label applied to end of box or wrapped package

Contd.

Table 12.2: Some of the common terms for labels *(Contd.)*

Neck label	Used for neck of bottle
Over-all wrap	Covers the entire surface of a container top, bottom and sides
Spot label	Label which covers only a small portion of the container
Tag	Special purpose label, Affixed to product or container by string wire, etc.
Wrap around label	Wrap all around the container, does not cover top or bottom

Pressure sensitive labels: Pressure sensitive labels may be considered as the most advanced form of labelling and is a process where the label is in a stage of permanent activation and does not require heat, moisture or gum in order to make it adhere to a surface. One can use finger pressure for sticking of the label on the surface. Paper sensitive label consists of a label paper coated with permanent tacky adhesive.

Swing labels: A tag may be described as a marking device that is attached to a container or product by some means other than adhesive—strings, ribbons, wire, holes and various types of slots and slits. High class food products, particularly those styled and designed as presentation and gift packages or units.

Transportation of Food Products

Transport is important because it enables trade between people, which is essential for the development of civilizations. Given the shifting trends in lifestyles and globalization of eating preferences, it has become imperative to devise ways to ensure availability of foods across the globe.

This can be achieved by developing methods of transporting food and food products safely to the consumers.

Transport or **transportation** is the movement of people, animals and goods from one location to another. Modes of transport include air, rail, road, water, cable, pipeline and space.

TRANSPORT MODES

o Air
o Land
  • Animal-powered
  • Human-powered
  • Rail
  • Road
o Cable
o Pipeline
o Space
o Water

COMPONENTS

Transportation encompasses the following in order to make its maximum impact:

1. **Infrastructure:** Transport infrastructure consists of the fixed installations including roads, railways, airways,

waterways, canals and pipelines as well as terminals such as airports, railway stations, bus stations, warehouses, trucking terminals, refueling depots (including fueling docks and fuel stations) and seaports. Terminals may be used both for interchange of passengers and cargo and for maintenance.

2. **Vehicles:** Vehicles traveling on these networks may include automobiles, bicycles, buses, trains, trucks, people, helicopters, watercraft, spacecraft and aircraft.

3. **Operations:** Operations deal with the way the vehicles are operated, and the procedures set for this purpose including financing, legalities and policies. In the transport industry, operations and ownership of infrastructure can be either public or private, depending on the country and mode.

MODES OF TRANSPORT

A mode of transport is a solution that makes use of a particular type of vehicle, infrastructure and operation. The transport of a person or of cargo/food may involve one mode or several of the modes, with the latter case being called intermodal or multimodal transport. Each mode has its own advantages and disadvantages, and will be chosen for a trip on the basis of cost, capability, and route.

Human-powered

Human-powered transport, a form of sustainable transportation, is the transport of people and/or goods using human muscle-power, in the form of walking, running and swimming. Modern technology has allowed machines to enhance human power. Human-powered transport remains popular for reasons of cost-saving, leisure, physical exercise, and environmentalism; it is sometimes the only type available, especially in underdeveloped or inaccessible regions.

Although humans are able to walk without infrastructure, the transport can be enhanced through the use of roads, especially when using the human power with vehicles, such as bicycles and inline skates. Human-powered vehicles have also been developed for difficult environments, such as snow and water, by watercraft rowing and skiing; even the air can be entered with human-powered aircraft.

Animal-powered

Animal-powered transport is the use of working animals for the movement of people and commodities. Humans may ride some of the animals directly, use them as pack animals for carrying goods, or harness them, alone or in teams, to pull sleds or wheeled vehicles.

Air

The aircraft is the second fastest method of transport, after the rocket. Commercial jets can reach up to 955 kilometres per hour (593 mph), single-engine aircraft 555 kilometres per hour (345 mph). Aviation is able to quickly transport people and limited amounts of cargo over longer distances, but incur high costs and energy use; for short distances or in inaccessible places helicopters can be used.

Rail

Rail transport is where a train runs along a set of two parallel steel rails, known as a railway or railroad. The rails are anchored perpendicular to ties (or sleepers) of timber, concrete or steel, to maintain a consistent distance apart, or gauge. The rails and perpendicular beams are placed on a foundation made of concrete, or compressed earth and gravel in a bed of ballast. Alternative methods include monorail and maglev.

A train consists of one or more connected vehicles that operate on the rails. Propulsion is commonly provided by a locomotive, that hauls a series of unpowered cars, that can carry passengers or freight. The locomotive can be powered by steam, diesel or by electricity supplied by trackside systems. Alternatively, some or all the cars can be powered, known as a multiple unit. Also, a train can be powered by horses, cables, gravity, pneumatics and gas turbines. Railed vehicles move with much less friction than rubber tires on paved roads, making trains more energy efficient, though not as efficient as ships.

Road

A road is an identifiable route, way or path between two or more places. Roads are typically smoothed, paved, or otherwise prepared to allow easy travel though they need not be, and historically many roads were simply recognizable routes without any formal construction or maintenance. In urban areas, roads may pass through a city or village and be named as streets, serving a dual function as urban space easement and route.

The most common road vehicle is the automobile; a wheeled passenger vehicle that carries its own motor. Other users of roads include buses, trucks, motorcycles, bicycles and pedestrians.

Road transport offers a complete freedom to road users to transfer the vehicle from one lane to the other and from one road to another according to the need and convenience. This flexibility of changes in location, direction, speed, and timings of travel is not available to other modes of transport. It is possible to provide door to door service only by road transport.

Automobiles provide high flexibility with low capacity, but require high energy and area use, and are the main source of noise and air pollution in cities; buses allow for more efficient travel at the cost of reduced flexibility. Road transport by truck is often the initial and final stage of freight transport.

Water

Water transport is movement by means of a watercraft—such as a barge, boat, ship or sailboat—over a body of water, such as a sea, ocean, lake, canal or river. The need for buoyancy is common to watercraft, making the hull a dominant aspect of its construction, maintenance and appearance.

Although slow, modern sea transport is a highly efficient method of transporting large quantities of goods. Transport by water is significantly less costly than air transport for transcontinental shipping; short sea shipping and ferries remain viable in coastal areas.

Other Modes

Pipeline transport sends goods through a pipe; most commonly liquid and gases are sent, but pneumatic tubes can also send solid capsules using compressed air. For liquids/gases, any chemically stable liquid or gas can be sent through a pipeline. Short-distance systems exist for sewage, slurry, water and beer, while long-distance networks are used for petroleum and natural gas.

Cable transport is a broad mode where vehicles are pulled by cables instead of an internal power source. It is most commonly used at steep gradient. Typical solutions include aerial tramway, elevators, escalator and ski lifts; some of these are also categorized as conveyor transport.

Spaceflight is transport out of Earth's atmosphere into outer space by means of a spacecraft. While large amounts of research have gone into technology, it is rarely used except to put satellites into orbit, and conduct scientific experiments. However, man has landed on the moon, and probes have been sent to all the planets of the Solar System.

COMPONENTS

Infrastructure is the fixed installations that allow a vehicle to operate. It consists of a way, a terminal and facilities for parking and maintenance. For rail, pipeline, road and cable transport, the entire way the vehicle travels must be built up. Air and water craft are able to avoid this, since the airway and seaway do not need to be built up. However, they require fixed infrastructure at terminals.

Terminals such as airports, ports and stations, are locations where passengers and freight can be transferred from one vehicle or mode to another. For passenger transport, terminals are integrating different modes to allow riders to interchange to take advantage of each mode's advantages. For instance, airport rail links connect airports to the city centers and suburbs. The terminals for automobiles are parking lots, while buses and coaches can operate from simple stops. For freight, terminals act as trans-shipment points, though some cargo is transported directly from the point of production to the point of use.

The financing of infrastructure can either be public or private. Transport is often a natural monopoly and a necessity for the public; roads, and in some countries railways and airports, are funded through taxation. New infrastructure projects can have high cost, and are often financed through debt. Many infrastructure owners therefore impose usage fees, such as landing fees at airports, or toll plazas on roads. Independent of this, authorities may impose taxes on the purchase or use of vehicles. Because of poor forecasting and overestimation of passenger numbers by planners, there is frequently a benefits shortfall for transport infrastructure projects.

Vehicles

A vehicle is a non-living device that is used to move people and goods. Unlike the infrastructure, the vehicle moves along with the cargo and riders. Unless being pulled/pushed by a cable or muscle-power, the vehicle must provide its own propulsion; this is most commonly done through a steam engine, combustion engine, electric motor, a jet engine or a rocket, though other means of propulsion also exist. Vehicles also need a system of converting the energy into movement; this is most commonly done through wheels, propellers and pressure.

Vehicles are most commonly staffed by a driver. However, some systems, such as people movers and some rapid transits, are fully automated. For passenger transport, the vehicle must have a compartment, seat, or platform for the passengers. Simple vehicles, such as automobiles, bicycles or simple aircraft, may have one of the passengers as a driver.

Operation

Private transport is only subject to the owner of the vehicle, who operates the vehicle themselves. For public transport and freight transport, operations are done through private enterprise or by governments. The infrastructure and vehicles may be owned and operated by the same company, or they may be operated by different entities. Traditionally, many countries have had a national airline and national railway. International shipping remains a highly competitive industry with little regulation, but ports can be public owned.

KINDS OF TRANSPORT

Short-haul transport is dominated by the automobile and mass transit. The latter consists of buses in rural and small cities, supplemented with commuter rail, trams and rapid transit in larger cities.

Long-haul transport involves the use of the automobile, trains, coaches and aircraft, the last of which have become predominantly used for the longest, including intercontinental, travel.

Intermodal passenger transport is where a journey is performed through the use of several modes of transport; since all human transport normally starts and ends with walking, all passenger transport can be considered intermodal.

Public transport may also involve the intermediate change of vehicle, within or across modes, at a transport hub, such as a bus or railway station.

Taxis and buses can be found on both ends of the public transport spectrum. Buses are the cheaper mode of transport but are not necessarily flexible, and taxis are very flexible but more expensive. In the middle is demand-responsive transport, offering flexibility whilst remaining affordable.

International travel may be restricted for some individuals due to legislation and visa requirements.

Freight or Shipping

Freight transport, or shipping, is a key in the value chain in manufacturing. With increased specialization and globalization, production is being located further away from consumption, rapidly increasing the demand for transport.

Transportation creates place utility by moving the goods from the place of production to the place of consumption. While all modes of transport are used for cargo transport, there is high differentiation between the nature of the cargo transport, in which mode is chosen. Logistics refers to the entire process of transferring products from producer to consumer, including storage, transport, trans-shipment, warehousing, material-handling and packaging, with associated exchange of information.

Containerization, with the standardization of ISO containers on all vehicles and at all ports, has revolutionized international and domestic trade, offering huge reduction in transshipment costs.

Bulk transport is common with cargo that can be handled roughly without deterioration; typical examples cereals and coal.

The low value of the cargo combined with high volume also means that economies of scale become essential in transport, and gigantic ships and whole trains are commonly used to transport bulk.

Air freight has become more common for products of high value; while less than one percent of world transport by volume is by airline, it amounts to forty percent of the value.

Time has become especially important in regard to principles such as postponement and just-in-time within the value chain, resulting in a high willingness to pay for quick delivery of key components or items of high value-to-weight ratio.

EVOLUTION OF TRANSPORTATION

First means of transport involved walking, running and swimming.

The domestication of animals introduced a new way to lay the burden of transport on more powerful creatures, allowing the hauling of heavier loads, or humans riding animals for greater speed and duration.

Inventions such as the wheel and the sled helped make animal transport more efficient through the introduction of vehicles.

Water transport, including rowed and sailed vessels, dates back to time immemorial, and was the only efficient way to transport large quantities or over large distances prior to the Industrial Revolution.

The Industrial Revolution in the 19th century saw a number of inventions fundamentally change transport. With telegraphy, communication became instant and independent of the transport of physical objects. The invention of the steam engine, closely followed by its application in rail transport, made land transport independent of human or animal muscles. Both speed and capacity increased rapidly, allowing specialization through manufacturing being located independently of natural resources. The 19th century also saw the development of the steam ship, which sped up global transport.

Thereafter the 20th and 21st century saw the transportation grow by leaps and bounds, given the necessity of modern day living.

Food commodities are a specialized item needing special care while transportation. Depending on the kind of food items, the modes of transportation are decided.

FOOD TRANSPORTATION

Modern food transport guarantees the amazing array of variety that exists in the supermarkets today. Huge refrigerated ships bring perishable fruits from tropical countries all over the world.

Evolution of Food Transportation

The earliest method to transport food was of course on foot, a method that is very limited. A big step forward was when packed animals, boats and barges started to come in use, it increased the capacity substantially. The invention of the wheel gave mobile carts; roads were built to make it faster and better for transport. These better transport methods made trade between people grow, pretty soon food and wares were shipped long distances.

Better roads, better ships and improved preservation methods gave way to greater food transportation.

These days there are almost no limits for what can be transported and at what distance. Food is preserved and packed in efficient ways, minimizing and simplifying transport. Frozen cargo can be transported by sea, land or air. Our modern society is depending on huge quantities of food stuffs to be transported.

FOOD TRANSPORTATION AND SAFETY

Since food transportation is a vital issue, special care needs to be taken to ensure that food transported is done in a safe and sanitary way.

Transportation of foods presents three types of hazards:

o **Physical hazards** such as pieces of metal, wood, glass or other foreign bodies which may find their way into foods while being transported.

o **Chemical hazards** from previous non-food cargoes, from non-food cargoes mixed in the same load; from refrigerant leaks, from residues of cleaning agents, or from the external environment.

o **Biological hazards** from contamination by bacteria, moulds, yeast, parasites, algae, and the growth of biological contaminants if the temperature control is inadequate.

Reducing or eliminating the above-mentioned hazards could be achieved through applying proper cleaning programmes and following good transportation practices (the most important of which is controlling the temperature during food transportation).

Good communication between shipper/manufacturer, transporter and receiver of foodstuffs is essential. They share responsibility for food safety on this part of the food chain. Food manufacturers or receivers are responsible for communicating to transporters specific food safety control procedures required during transportation. Hence all food control systems have put a lot of emphasis on food transportation and indicated that it could be detrimental in preventing many health and economic catastrophes for the consumers and food manufacturers.

The two main food safety issues are

o Keeping the food protected from contamination

o If the food is potentially hazardous, keeping it cold (5°C or colder) or hot (60°C or hotter).

Protecting Food from Contamination

It is important to protect food from contamination by keeping it covered at all times. This can be achieved by using containers with lids or by applying plastic film over containers. Materials used to cover food should be suitable for food contact, to ensure that they do not contain any chemicals that could leach into the food. Aluminium foil, plastic film and clean paper may be used, and food should be completely covered. Packaged products should not need additional covering.

Previously used materials and newspaper may contaminate food and should not be used.

Temperature Control

When potentially hazardous foods are transported they should be kept cold (5°C or colder) or hot (60°C or hotter) during the journey. Alternatively, you could use time, rather than temperature, to keep the food safe while it is being transported.

If the journey is short, insulated containers may keep the food cold. If the journey is longer, ice bricks to keep food cold and heat packs to keep food hot will have to be used.

Place only pre-heated or pre-cooled food in an insulated container, which should have a lid to help maintain safe temperatures.

Insulated containers must be:

o in good condition and kept clean at all times
o used only for food
o kept away from other items such as chemicals, pet food, fuel and paint
o be filled as quickly as possible and closed as soon as they have been filled.
o kept closed until immediately before the food is needed or is placed in other temperature-controlled equipment.

Some points to note therefore are discussed below:

Food carriers

Carriers used by the manufacturer are designed, constructed, maintained, cleaned and utilized in a manner that prevents food contamination and minimizes microbial growth.

o Carriers, including bulk tanks, are clean, dry, weatherproof, free of infestation and sealed to prevent water, rodents or insects from reaching the products.

o The carrier has an adequate cleaning and sanitizing program in place. All tankers delivering raw ingredients, such as flour, are thoroughly cleaned on a set schedule and each time there is a possibility of cross-contamination with allergens between shipments, in a cleaning station before being loaded. The adequacy of cleaning is verified.

o Where direct contact with food may occur, materials used in carrier construction are suitable for food contact.

o Conveyances and containers are inspected by the manufacturer upon receipt of incoming ingredients and prior to loading of final products to ensure they are free from contamination (e.g. pests, residues) and suitable for the transportation of food.

o Proper environmental conditions such as temperature and humidity are controlled, monitored, and documented to assure raw material and finished product safety and wholesomeness.

o Where the same carriers are used for food and non-food loads, procedures are in place to restrict the type of non-food loads to those that do not pose a risk to food loads in the same shipment, or to subsequent food loads after an acceptable clean out.

o Fresh alimentary paste products are loaded, transported and unloaded in a manner that protects them from any damage and/or contamination.

o Bulk tanks are designed and constructed to permit complete drainage and to prevent contamination.

o Carriers are loaded, arranged and unloaded in manner that prevents damage and/or contamination of the food.

Temperature Controls

Ingredients and finished product requiring temperature controls are transported in a manner to prevent temperature abuse that could result in deterioration of the product and affect its safety.

o Ingredients requiring refrigeration are transported at 4°C (39°F) or less and the temperature is appropriately monitored.

○ Frozen ingredients are transported at temperatures that do not permit thawing and the temperature is appropriately monitored.

○ Finished product is transported under conditions that minimize microbiological, physical and chemical deterioration.

Table 13.1: Transport temperature requirements of food products

Chilled products	Temperature (°C)
Fresh fish (in ice), crustaceans and shellfish (excluding live ones)	+2
Cooked dishes and prepared foods, pastry creams, fresh pastries, sweet dishes and egg products	+3
Meat and cooked meats pre-packaged for consumer use	+3
Offal	+3
Poultry, rabbit	+4
Non-sterilized, untreated, unpasteurised or fermented milk, fresh cream, cottage cheese and curd	+3
Milk for industrial processing	+6
Cooked meats other than those which have been salted, smoked, dried or sterilized	+6

Frozen products	Temperature (°C)
Ice and ice cream	−25
Deep frozen foods	−18
Fishery products	−18
Butter and edible fats, including cream to be used for butter making	−14
Egg products, offal, rabbit, poultry	−12
Meat	−10

Transport Considerations

○ Containers of cool food should be placed in the coolest part of the vehicle.

○ If the inside of the vehicle is air-conditioned, cold food may be transported better here rather than in the boot.

o Vehicles should be clean. If the vehicle is normally used for carrying pets or dirty equipment, the food carrying area should be thoroughly cleaned or lined to prevent any contamination. This may not be necessary if food is transported in an insulated container with a tightly fitting lid.

o The journey should be properly planned and should be kept as short as possible.

o When collecting ingredients, cold foods should be collected last and immediately placed in insulated containers or cool bags for transporting to the preparation facility.

o When taking prepared foods to a venue, pack the food into insulated boxes in the end.

o On reaching the venue it is a priority to unload any hot or cold food and place it in temperature-controlled equipment.

Storage

Incoming Materials Storage

Storage and handling of incoming ingredients and packaging materials is controlled to prevent damage and contamination.

o Ingredients requiring refrigeration are stored at 4°C (39°F) or less and the temperatures are appropriately monitored.

o Frozen ingredients are stored at temperatures that do not permit thawing and the temperatures are appropriately monitored.

o Humidity sensitive ingredients and packaging materials are stored under appropriate conditions to prevent deterioration.

o Incoming materials and packaging materials are handled and stored in a manner that prevents damage and/or contamination (including cross-contamination with allergens).

o Incoming materials containing allergens are clearly identified (using a colour code or other identification system).

o Incoming materials and packaging materials are stored off the floor and away from walls to permit access for cleaning and pest control.

o All raw materials are kept separate from packaging material or finished product.

○ Stock rotation (of ingredients and where appropriate, packaging materials) is controlled to prevent deterioration and spoilage (e.g. first-in, first-out).
○ Defective or suspect product or ingredients are clearly identified and isolated in a designated area for appropriate disposition.

Non-food Chemicals—Receiving and Storage

Non-food chemicals are received and stored in a manner that prevents contamination of food, packaging materials and food contact surfaces.

○ Non-food chemicals are received and stored in a dry, well ventilated area.
○ Non-food chemicals are stored in designated areas such that there is no possibility for cross-contamination of food or food contact surfaces.
○ Where required for ongoing use in food handling areas (e.g. conveyor lubricants), these chemicals are stored separate from food and located in a manner that prevents contamination of food, food contact surfaces or packaging materials.
○ Non-food chemicals are stored and mixed in clean, correctly labelled containers that include instructions for use.
○ Non-food chemicals are dispensed and handled only by authorized and properly trained personnel.

Finished Product Storage

Finished product is stored and handled under conditions that prevent damage and contamination.

○ Fresh alimentary paste products are stored and handled under conditions that minimize damage, deterioration and prevent contamination including allergen contamination.
○ Finished products are stored off the floor and away from walls to permit access for cleaning and pest control.
○ Finished products requiring refrigeration are stored at 4°C (39°F) or less and are appropriately monitored.
○ Frozen finished products are stored at a temperature that does not permit thawing.

o Stock rotation is controlled to prevent deterioration and prevent spoilage of products that could present a health hazard (e.g. products exceeding shelf life).

o Returned, defective or suspect product is clearly identified and isolated in a designated area for appropriate disposition.

o Finished product is stored and handled in a manner that minimizes damage (e.g. forklift damage or damage due to uncontrolled stacking heights).

Transport Refrigeration Technologies

A transport refrigerator vehicle is designed to carry perishable freight at specific temperatures. Like refrigerator cars, refrigerated trucks differ from simple insulated and ventilated vans (commonly used for transporting fruit), neither of which are fitted with cooling apparatus.

Refrigerator vehicle can be ice-cooled, equipped with any one of a variety of mechanical refrigeration systems powered by small displacement diesel engines, or utilize carbon dioxide (either as dry ice or in liquid form) as a cooling agent.

Different transport techniques are as follows:

o **Water cooling:** Water cooling systems are expensive, so modern vessels rely more on ventilation to remove heat from cargo holds, and the use of water cooling systems is declining. Air cooling and water cooling are usually combined. Air cooling removes the heat generated by the reefers while water cooling helps to minimize the heat rejected by the reefers. The reefers are using some heat exchangers that behave as water cooled condensers.

o **Cryogenic cooling:** Another refrigeration system sometimes used where the journey time is short is total loss refrigeration, in which frozen carbon dioxide ice (or sometimes liquid nitrogen) is used for cooling. The cryogenically frozen gas slowly evaporates, and thus cools the container and is vented from it. The container is cooled for as long as there is frozen gas available in the system. These have been used in railcars for many years, providing up to 17 days temperature regulation.

Whilst refrigerated containers are not common for air transport, total loss dry ice systems are usually used.

These containers have a chamber which is loaded with solid carbon dioxide and the temperature is regulated by a thermostatically controlled electric fan, and the air freight versions are intended to maintain temperature for up to around 100 hours. Full size intermodal containers equipped with these "cryogenic" systems can maintain their temperature for the 30 days needed for sea transport. Since they do not require an external power supply, cryogenically refrigerated containers can be stored anywhere on any vessel that can accommodate "dry" (un-refrigerated) ocean freight containers.

o **Redundant refrigeration:** Valuable, temperature-sensitive, or hazardous cargo often require the utmost in system reliability. This type of reliability can only be achieved through the installation of a redundant refrigeration system. A redundant refrigeration system consists of integrated primary and back-up refrigeration units, i.e. a container fitted with two refrigeration units and a single diesel generator. If the primary unit malfunctions, the secondary unit automatically starts. To provide reliable power to the refrigeration units, these containers are often fitted with one or more diesel generator sets.

o **Mechanical refrigeration:** All mechanical transport refrigeration units normally includes a compressor, drive, and condenser combination; an evaporator or air-cooler; all necessary refrigerant lines and electrical wiring; and means whereby the unit can be suitably mounted and installed on a vehicle, used in transportation of perishable goods. The first successful mechanically refrigerated trucks were introduced by the ice cream industry in about 1925. There were around 4 million refrigerated road vehicles in use in 2010 worldwide.

The development of the transport refrigeration unit using the latest state-of-the-art technology, which sets the benchmark for the premium segment requires:

o Precise temperature control throughout the entire interior with minimal fluctuation.

o Excellent economy combined with high-performance cooling through efficient motor management.

o High refrigeration performance for fast cooling down.

o The best heat output among the direct competitors for the shortest possible interruptions in goods cooling when defrosting.

o 50% fewer defrosting cycles due to the ice-reducing evaporator design.

o Durable industrial motor with intelligent speed control and compressor with cylinder deactivation for low fuel consumption.

o Greater operational reliability and control thanks to the electronic control unit.

o Longer maintenance intervals due to the specially dimensioned wearing parts.

o Transport refrigeration systems designed specifically for make and model of vehicle.

o The size of all components is carefully calculated to give maximum cooling output without adversely affecting the vehicle's fuel consumption for performance.

o To maintain the system in good working order and to minimize the loss of refrigerant, the refrigeration unit should be operated for a minimum of five minutes each week regardless of the season. This will assist in preventing the compressor seal from drying out; a condition which can cause loss of refrigerant and possible damage to the compressor.

Environmental Considerations

For the first time, the prestigious competition recognized the urgent need to develop sustainable transport refrigeration systems to replace the highly polluting diesel-powered Transport Refrigeration Units (TRUs) that dominate the industry today. Companies are developing innovative cooling refrigeration technologies to reduce greenhouse emissions and local air pollution.

Transport refrigeration requires a high level of system performance in some of the most demanding environments imaginable with reliability, precision temperature control and efficiency. Ten years ago, less than 5% of the refrigerated container users employed scroll compressors. Now over half of the users employ scroll. Truck refrigeration and rail air conditioning have also seen a significant migration to scroll compressors. Need to produce the transport refrigeration system that is engineered entirely without compromise utilizing the ground-breaking, liquid nitrogen powered engine, the system is being developed to offer industry leading performance, zero-emission and quiet operations, all without having to compromise on cost.

Transport refrigeration is not a one-size-fits-all market. Rather, it is one in which environmental challenges and regulatory changes are rapidly driving the development of technologies specially adapted to provide maximum flexibility. To keep the transport industry moving forward efficiently and the food chain developing safely, it is critical that refrigeration component and equipment manufacturers establish close partnerships to develop solutions that help the industry meet its evolving performance objectives while anticipating future demands and challenges.

Chapter 14

Sensory Evaluation and Quality Control

"Every time food is eaten a judgment is made", said Jerold, adding, "Quality and not quantity is my ultimate criterion". Indeed, quality is very important parameter for judging the edible nature of any food.

When the quality of a food product is assessed by means of human sensory organs, the evaluation is said to be sensory or subjective or organoleptic.

Sensory quality is a combination of different perceptions that come into play in choosing and eating food. Appearance, which can be judged by the eye, e.g. colour, size, shape, uniformity and absence of defects, is of first importance in food selection.

Sensory evaluation *is a scientific discipline used to evoke, measure, analyze and interpret reactions to those characteristics of foods and material as they are perceived by the senses of sight, smell, taste, touch and hearing.*

The effective characteristic is not the property of the food, but the subject's reaction to the sensory qualities of food. This reaction is highly conditioned by a variety of psychological and social factors and in the final analysis, plays a vital role in the acceptance and preferences of foods.

SENSORY CHARACTERISTICS OF FOOD

Appearance

Surface characteristics of food products contribute to the appearance.

A biscuit with very dry surface is not acceptable, whereas biscuits with crispy surface are rated high.

The size and shape of pieces of food and the brown colour of crusts of biscuits are judged by eye. Sight also plays a role in the assessment of the lightness of biscuits. Completeness of baked biscuits can be judged by appearance of the products.

Colour

Colour is used as an index of the quality of the product. Biscuits, which are too brown, are likely to be rejected in anticipation of scorched bitter taste.

Flavour

The flavour of food has three components—odour, taste and composite of sensations known as mouthfeel.

o *Odour:* The odour of food contributes to the pleasure of eating. A substance, which produces odour, must be volatile and the molecules of the substance must come in contact with receptors in the epithelium of the olfactory organ. The volatility of aromas is related to the temperature of the food. High temperature tends to volatalize aromatic compounds and help in judging the state of the food, whereas cool or cold temperature inhibit volatalization.

o *Taste:* We value food for its taste. Taste sensations, which the taste buds register, are categorized as sweet, salty, sour or bitter. Taste buds in the different areas of the tongue are not equally sensitive to all taste stimuli and at least some taste cells respond to more than one stimulus. Taste buds near the tip of the tongue are more sensitive to sweet and salt. Salty taste is due to ions of salt. Sugars are the main source of sweetness in food. Not all sugars are equally sweet. Fructose is the sweetest. The concentration required for identification is known as the "threshold". Threshold for each of the primary tastes is usually not at the same level in any one individual.

o *Mouthfeel:* Texture and consistency can be found out by mouthfeel.

Texture

It can be characterized by how the food feels on the tongue — coarse or fine. Coarse textured crystalline products are said to be grainy. The brittleness of food is another aspect of texture. Good quality biscuits are crispy.

PSYCHOLOGICAL FACTORS

In addition to colour, odour, taste and mouthfeel, certain psychological factors contribute to the acceptability of food. Food is accepted when there is a pleasant association.

USES

- Evaluate a range of existing food products
- Analyze a test kitchen sample for improvements
- Gauge consumer response to a product, and
- Check that the final product meets its original specifications.
- Food product development entails that companies or individuals want to
- Develop new products by modifying existing formulations
- Enter new markets
- Compete more effectively in existing market, and
- Keep a high level of quality.

These intentions, therefore, make sensory evaluation very essential.

Food quality, or from the consumer viewpoint, food acceptance is the most critical aspect of food. The collection of food acceptance data is a key component in studies on product development, quality control, food product acceptance in the market place and food service evaluation.

Usually, a sensory panel is constituted to conduct periodic quality assessment of the product, followed by consumer testing.

SENSORY PANEL

During the product development cycle, it is necessary to conduct periodic quality assessment of the product being developed. This is done either with a consumer panel or a trained panel.

The trained panel is selected and trained in such a way that the panelists are capable of giving high reliability of judgements independent of psychological factors such as bias, motivation and individual experience. The expert is not viewed as representing the consumer. The role of the expert in product development is to determine flaws in the development process (too salty, poor texture, etc), and possibly to attribute these flaws to specific processing steps, e.g. burnt note attributable to overheating or higher processing temperature. This information is used by the developer to alter the formulation or to improve processing. As a result of experimentation and sensory evaluation, an optimized product is developed. This is then submitted to a consumer to be maximally effective.

Consumer Testing

Consumer testing is the next crucial aspect in product development. Consumer testing can be done in three ways:
1. in-house laboratory testing,
2. home testing, and
3. institutional testing
1. *In-house laboratory* acceptance testing represents the most controlled environment in which to conduct acceptance tests. Within the laboratory testing area, one can control a number of environmental variables (odour, light, temperature, humidity, etc.) and a number of stimulus variables (serving temperatures, portion size, etc.). In-house testing utilizes either laboratory personnel or consumers brought in for the tests.
2. *In home testing*, the selection and maintenance of a consumer panel is a key issue. The cooperation rate from consumer home panels is approximately 50 per cent. It has been found that cooperation is best in households with:
   a. more than two members,
   b. a younger housewife, and
   c. more education.
   But home testing presents a practical problem, i.e. the process of data collection is not done under the supervision of the investigator. Therefore, validity of the procedure and resulting data cannot be directly assessed.

3. The *institutional food service setting*, on the other hand, provides an excellent opportunity to collect food acceptance information. It is preferable to collect the food acceptance ratings from the consumers as they are eating or just after they have completed eating. Collection of direct consumer acceptance ratings also provides the researcher with an opportunity to observe the food system in operation and to interact with consumers.

Food acceptance data is collected by using **feedback forms** that are filled by the consumers. For their effective use, the feedback forms

- o should be brief and clear, it should take only 1–2 minutes to fill out.
- o the format of questions should provide information which can be acted upon.

Sensory Evaluation during Product Life Cycle

The basic procedure for developing a new product and supporting it while marketed, included distinct steps that are constant no matter what type of product is produced.

Initial screening in product development roughly defines the final product. The objective during this phase of life cycle is to formulate and physically prepare a prototype that is close to the final product, yet knowing the product will go through extensive optimization. The different stages during product life cycle are summarized herewith:

a. **Product optimization:** Sensory analysis during this phase of product development is critical and includes extensive evaluation with many kinds of tests, each playing a specific role in optimization of the new product. Trained descriptive panels are used to characterize the flavour profile and other characteristics compared to what is already in the market. Panels are used to determine product acceptability and aid in defining the formula and product specifications such as moisture, oil, salt, seasoning and oil flavour in fresh and aged products. Product testing by consumer panel can be conducted by the company developing the product or by an independent consumer evaluation agency.

b. **Scale up:** At this phase, sensory analysis consists of tests that compare the production samples with the optimized product. Depending on the resources available, either consumer panels or descriptive panels can qualify the production samples.

Sensory specifications are also determined before the product is taken into full production. This is a time-consuming process, similar to establishment of analytical specifications. The first step consists of screening of samples that represent reasonable extremes in the manufacturing process and also represent different raw material samples. Descriptive analysis is then used to characterize the products in quantitative terms. Consumer data are used to determine which attributes are critical and to set acceptable limits around the optimum target.

c. **Production:** Sensory analysis does not stop after the product has been developed and is being produced routinely. However, it is critical that products continue to be analyzed to ensure the finished goods are consistently manufactured to design criteria and that the product profile does not "drift" over time.

Typically, products and packages are inspected shortly after production in what is sometimes called a sample-cutting meeting. Persons involved in evaluating freshly made products must become familiar with how products with varying characteristics age during their expected shelf-life. Traditional difference and/or variation testing should be conducted on a routine basis for quality assurance purposes. At this point, shelf-life testing should be conducted to ensure the product meets specifications till the end of its declared shelf-life.

Sensory tests are well integrated with the overall plan of development of the product. The sensory test plan given below is followed in detail before the samples are prepared and packed.

1. Definition of the specific objectives or questions required to be answered by the tests.

2. The selection of appropriate method.
3. The number of evaluations.
4. Statistical design to decide the quantity of sample to be prepared.
5. Statistical treatment of the data.

Preliminary experimentation is required to decide the optimum state, quantity and presentation of samples. Homogeneity, appropriateness and randomization are achieved to take care of different types of bias that could affect the judgements of personnel.

The method available for sensory analysis of foods can be broadly grouped as follows:

a. Difference tests
b. Rating tests
c. Sensitivity tests
d. Descriptive tests

The selection of a particular test will depend on the defined objective of the test, accuracy desired and personnel available for conducting the evaluation.

A. DIFFERENCE TESTS

A1. Paired Comparison Test

The samples to be tested and evaluated are given in pairs to the panelists. The pairs can either be of the same samples or different samples. To avoid prejudices the samples are presented in code numbers. The panelist has to rate the samples according to a distinguishing characteristic. The samples are different in the intensity of that characteristic and so the more desirable one has to be rated higher.

Specimen Evaluation Card

Paired Comparison Test

Name: .... Test: ....
Product: ....

You are given one or several pairs of samples. Evaluate the two samples in the pair for texture, colour, etc.

Is there any difference between the two samples in the pair?

Code No. of pairs	Yes	No
----------- -----------	-----------	-----------
----------- -----------	-----------	-----------
----------- -----------	-----------	-----------

Signature

A2. Duo-trio Test

As the name suggests this test comprises three samples, two of which form a group are identical while the third is different. At first the panelist is presented with one of the identical samples. This is the reference sample. Thereafter, the other two successively in random order, are given for testing. The panelist is required to match one of them with the reference. A positive answer is mandatory even if a guess is made. The chance probability of placing the samples in a certain order is one-half.

Specimen Evaluation Card

Duo-trio Test

Name: ....   Date: ....

Product: ....

The first sample 'A' given is the reference sample. Taste it carefully.

From the pair of coded samples next given, judge which sample is the same as 'A'. A positive answer is to be made even if it is a guess.

	Set no.	Code no. of pairs	Same as 'A'
I.	-------------	-----------	-----------
II.	-------------	-----------	-----------
III.	-------------	-----------	-----------
IV.	-------------	-----------	-----------

Signature

A3. Triangle Test

Similar to the Duo-trio test, this has a slight difference. Although, even here three samples are used, one different two identical: The panelist is presented with all the three samples simultaneously. The panelist is asked to identify the odd sample. A positive answer is required even if it is a guess. Since all three samples are unknown, the chance probability of placing the sample in a certain order is one-third.

Specimen Evaluation Card

Triangle Test

Name: ....   Date: ....

Product: ....

1. Here are three samples for evaluation. Two of the three samples are identical.
   Determine the odd sample for any difference.

   Sample*   Check odd
   AAB --------
   ABA --------
   BAA --------
   Did you check by guess?   Yes ☐   No ☐

2. Indicate the degree of difference between like and odd samples.
   None ....   Slight ....
   Moderate ....   Much ....

3. Acceptability
   Odd sample preferred   Yes....................   No.................
   Like sample preferred   Yes....................   No.................

4. Comments, if any.
   *In the actual test, use three figure numbers.

B. RATING TESTS

These tests give more quantitative data than difference tests and can be used for the analysis of more than two samples at the same time.

B1. Ranking Test

This test requires the panel to rank the samples on the basis of a characteristic. No pairs or reference samples are needed. As many samples as require are presented to the panelists are they are asked to rank them according to one or various characteristics. Samples as always are coded. This test is useful when conducting consumer preference analysis, as this clearly tells of the most favoured product.

Specimen Evaluation Card

Ranking Test

Name: ....   Test: ....

Product: ....

Please rank the samples in numerical order according to your preference or intensity of aroma/taste characteristic of the product.

Intensity/Preference	Sample code
First	------------
Second	------------
Third	------------
Fourth	------------

Comments: (Type of off-flavour, etc.)

B2. Single Sample (Monadic) Test

This test is advocated for samples of foods which leave an after taste or flavour which remains for some time. This inhibits the testing of some other sample at the same session due to fear of interference and mixing of flavours. The panelists are requested to point out the presence or absence and/or intensity of specified quality characteristics. The analysis of multiple samples can be made by comparing the test results of two or more samples evaluated at different times by trained panelists.

Specimen Evaluation Card

Single Sample (Monadic) Test

Name: ....   Date: ....

Product: ....

Please sniff and taste the sample carefully. Can you detect any off-flavour in the product ?

Circle one   Yes/No

If you detect any off-flavour, please describe it below:

Intensity (circle one)	Comments
Trace	Off-flavour is due to —
Moderate	Off-odour
Strong	Off-taste
	Residual taste
	Other defects

Signature

B3. Two-sample Difference Test

Panelists are served four pairs of samples. The pairs are so made that one of them is a reference sample while the other is a coded test sample.

Two pairs have both samples as reference samples while other two have one of each, i.e. reference as well as test sample.

The samples are then tested according to the Evaluation Card.

Specimen Evaluation Card

Two Sample Difference Test

Name: ....   Date: ....

Product: ....

1. Compare the coded sample to the reference sample independently in each of the four pairs given. Test sample may or may not be different from the reference sample.
2. Determine the degree and direction of difference on the following scale.

Degree		Direction	
No difference	- 0	Superior to standard	- S
Very slight difference	- 1	Equal to standard	- E
Moderate difference	- 2	Inferior to standard	- I

3. Comment on what the difference is based on:
   Odour, taste or both

Sample code No.	Degree of difference	Direction	Comment
--------	----------	--------	---------
--------	----------	--------	---------
--------	----------	--------	---------
--------	----------	--------	---------

(*Note*: If there is no difference, there is no degree or direction)

Signature

B4. Multiple Sample Difference Test

This test has the capability to evaluate more than one test variable. The reliability of the analysis is reduced though. The panelists are served three to six samples depending upon the number of test variables. Out of these, one is a standard reference sample. The panelist compares each coded sample with a known standard. One coded sample is a duplicate of the sample. Scores are then assigned to the samples by the panelists after evaluation.

Specimen Evaluation Card

Multiple Sample Difference Test

Name: .... Date: ....

Product: ....

You are given a standard or reference sample marked 'R'. Taste it carefully for the quality characters to be evaluated. You are next given a number of samples which are to be compared to the reference sample for odor and flavour. Rate each sample in degree of difference and the direction of quality according to the following scale:

Rating	Degree of direction Difference from standard		Direction of quality
0	None	E	Equal
1	Slight	I	Inferior
2	Moderate	S	Superior
3	Large		

Sample Code No.	Odour Degree of Direction Comments	Flavour Degree of direction Comments
-----	------	
-----	------	
-----	------	

Signature

B5. Hedonic Rating Test

These tests are aimed at measuring how acceptable food product is vis-à-vis its unacceptability. This test rates consumer acceptability. A food product is rated according to how much it is liked or disliked by the panelist. The limits of the scale, usually from 0 to 9, are from "extremely like" to "extremely dislike". After the rating on pre-mentioned scales, the results are analyzed for preference with data from large untrained panel.

In order to shortlist products, semi-trained panels are used to screen a large number of products so that only a few are left for consumer analysis.

If testing of a single product is to be made, separate evaluation cards are used while for testing and comparing multiple samples a single evaluation card with equal number of columns is used.

Specimen Evaluation Card

Hedonic Rating Test

Name: ....   Date: ....

Product: ....

     Taste these samples and check how much you like or dislike each one. Use the appropriate scale to show your attitude by checking at the point that best describes your feeling about the sample. Please give a reason for this attitude. Remember, you are the only one who can tell what you like. An honest expression of your personal feeling will help us.

1	Code:	Code:	Code:
Like extremely	----------	----------	----------
Like very much	----------	----------	----------
Like moderately	----------	----------	----------
Like slightly	----------	----------	----------
Neither like nor dislike	----------	----------	----------
Dislike slightly	----------	----------	----------
Dislike moderately	----------	----------	----------
Dislike very much	----------	----------	----------
Dislike extremely	----------	----------	----------
Reason	----------	----------	----------

Signature

B6. Numerical Scoring Test

This rating is done on the basis of a numerical score. The different samples to be evaluated are presented to the panelists and evaluation done. The samples are coded and are unknown to the panelists. The panelists assign scores to the samples on the basis of their like and dislike. Sensory characteristics corresponding to the agreed quality descriptions and scores is the basis of training of the panelists. This understanding of rating is very important.

Specimen Evaluation Card

Numerical Scoring Test

Name: ....   Date: ....

Product: ....

    Please rate these samples according to the following descriptions:

Score	Quality description
90	Excellent
80	Good
70	Fair
60	Poor

Sample	Score	Comments

Signature

B7. Composite Scoring Test

In this rating scale, each characteristic is rated separately and then the scores are compounded. The rating scale is then weighed so that the most important characteristic accounts for a large portion of the total score. The compounding of scores is done for one panelist. The advantage of the method and rating scale is that one can easily find out which attribute is poor and which is the best. This analysis is helpful in making changes in food products in specific characteristics of taste, odour, smell, etc.

Specimen Evaluation Card

Composite Scoring Test

Name: ....   Date: ....

Product: ....

Quality	Possible score	Sample scores			
Colour	20	—	—	—	—
Consistency	20	—	—	—	—
Flavour	40	—	—	—	—
Absence of defects	20	—	—	—	—
Total score	100				

Comments :

Signature

C. SENSITIVITY TESTS

As the name suggests these tests are aimed at assessing the ability of people to identify and detect different aspects of food quality in the presence of specific factors. The quality of food is assessed on the basis of taste, flavour, odour, feel and texture. The tests are targeted at identification of basic tastes like sweet, sour, salty, acidic and bitter. Thereafter, the assessment of the degree and concentration of the specific tastes has to be established. These tests help in training panel members for evaluation of food quality of products with spices, salt and sugar, e.g. sauces, ketchups, etc.

C1. Sensitivity-threshold test

In the field of food product development these tests are very useful. They measure the ability of an individual to detect specific parameters by smelling, tasting or feeling the food and making judgments. The tasters test food, beverages or even pure substances for specific characteristics. The intensity of sensory response to food or its components is established by these tests.

Specimen Evaluation Card

Sensitivity-threshold Test

Name: .... Date: ....

You receive a series of beakers with increasing concentrations of one of the four taste qualities (sweet, salty, sour, bitter). Start with beaker No. 1 and continue with beaker No. 2, No. 3, etc. Retasting of already tasted solutions is not allowed. Describe the taste or give intensity scores.

Use the following intensity scale

? = Different from water, but taste quality not identifiable

O = None or the taste of pure water

1 = Weak

2 = Medium

3 = Strong

4 = Very strong

5 = Extremely strong

)(Threshold very weak (taste identifiable)

Set No.	Description of taste and feeling factors
1	....
2	....
3	....
4	....
5	....
6	....
7	....
8	....
9	....
10	....
11	....
12	....

Signature

Threshold test: The aim of these tests is the measurement of threshold. Threshold is a point or level at which a transition of sensation or judgment occurs. It is a statistically determined point. There is a change from either no sensation to sensation or a level of sensation.

Basically three kinds of thresholds have been identified:

Stimulus detection threshold	Recognition identification threshold	Terminal saturation threshold
Magnitude of stimulus at which a transition occurs from no sensation to sensation.	A minimum concentration at which a stimulus is identified and acknowledged	It is the level of stimulus beyond which no apparent increase is recorded or perceived.

C2. Dilution Test

These tests are designed to identify and detect the presence of an unknown substance added in small amount to a standard product and developed as its substitute. It is detected when it has been added to the standard product to see how easily it mixes and blends with it.

Tests are employed for margarine in butter, dried whole milk in fresh milk, synthetic flavours with natural ones, synthetic essence with the natural essence, chemical spice with natural spice, etc.

The test material's quality and acceptability is represented by a dilution number. Dilution number is the percentage of the test material in the mixture of the standard product.

The dilution number represents presence of test material in such amount that there is an identifiable difference in characteristic of odour and taste between the test material and the standard product.

The relation that exists between the dilution and food quality is such that "bigger the dilution number better is the quality of the test material".

D. DESCRIPTIVE PROFILE METHOD

This test is so designed that it assesses both the qualitative and quantitative aspects of food or products. It tests the flavour in products with different tastes and odours. An example has been given below:

Flavour Analysis of Tomato Ketchup

Aroma: This is due to the various ingredients in ketchup which are:

Garlic	1
Cloves	1
Pepper	2
Cinnamon	2
Onion	3

This indicates that onion has the most distinct and dominating aroma.

Taste: The taste of ketchup is a combination of sour sweet and salty. But the sweet taste contributed by sugar dominates the other two and so it appears sweeter than salty.

Sugar (sweet)	2
Tomatoes (sour)	1
Salt	1
Mouthfeel	
Chillies: They impart a tangy feel to the ketchup	1

Texture: Since all the ingredients are blended and cooked well and are also strained to avoid any fibrous substance, the texture is smooth.

Smoothness	3

Method	Panelists		No. of samples
	Type	Number	per cent
A. Difference (qualitative)			
1. Paired comparison	Trained	5–12	2
	Untrained	72–80	
2. Duo-trio	Trained	5–12	3 (2 identical and 1 different)
3. Triangle (triad)	Trained	5–12	3 (2 identical and 1 different)
B. Rating (quantitative differences)	Trained	5–12	2–7

Contd.

Method	Panelists		No. of samples per cent
	Type	Number	
1. Ranking	Semi-trained	10–25	
	Untrained	72–80	
2. Single sample (Monadic)	Trained	6–25	1
	Untrained	72–80	
3. Two-sample difference	Trained	6–25	4 pairs of unknown and control sample
4. Multiple sample and quality difference	Trained	6–25	3–6
	Semi-trained	10–25	Including control and depending on number of quality factors evaluated
5. Hedonic	Semi-trained	10–25	Larger number only if mild flavoured or rated for colour or texture
6. Numerical scoring	Untrained	72–80	1–4
	Trained	5–12	5–10
7. Composite	Trained	5–12	1–4
C. **Sensitivity**			
1. Threshold	Untrained		
2. Dilution	Trained	12–14	5–10
D. **Descriptive Flavour Profile**	Trained (specially in the technique)	3–6	1–5

Sensory Evaluation Methods

Statistical analysis of data	Purpose
A.	
1. Binomial distribution x^2-test	Comparing samples for specific characteristics
2. Binomial distribution	Detecting difference when carryover, after-taste may be present. Also for training and testing panels.
3. Binomial distribution	Detecting differences when inter sample effects (after-taste, etc.)

Contd.

Statistical analysis of data	Purpose
B.	
1. Rank analysis	Determining order according to one specific characteristic or determining preference used in product and process analysis. Consumer perference analysis.
2. Analysis of variance	Detecting difference from normal product, off flavour, off taste and direction when after-taste and carryover are consumer analysis.
3. Analysis of variance	Difference between samples, quantitatively and directionally.
4. Analysis of variance	Comparing samples with more than one variable in the same session (reduced reliability).
5. Analysis of variance	Pilot consumer analysis for screening by preference.
6. Analysis of variance	Consumer analysis for preference. Screening for quality new product development, quality maintenance, ingredient or product grading and selection of trained panel.
7. Analysis of variance	Comparing several products of same type and grading.
C.	
1. Threshold value	Selecting panel members for evaluating ingredients, packaging material; maintaining quality.
2. Dilution number	Odour and flavour evolution of foods; ingredients; product development; quality control, specially useful for spices.
D.	Sample characteristics expressed in common terms. Intensity expressed on agreed scale. Used in new product development, product improvement and storage studies.

REQUIREMENTS FOR CONDUCTING SENSORY TESTS

- **Trained panel members:** The sensory qualities, particularly the flavour, are essentially measured subjectively. The trained panel generally comprises a small number of people who, in a rigorously controlled set-up of the laboratory, look after quality control of in-line and final product, process development and, to a limited extent, preliminary acceptance testing. Periodically, the panel is given refresher training and tests. Flavour profile and texture profile panels should be trained to an even greater degree since there is detailed qualitative and quantitative analysis. The trained panels are of two kinds:

a. **Discriminative, communicative or semi-trained panels:** They comprise technical people and their families. They are generally familiar with qualities of various foods. Due to their experience they are capable of giving impartial and accurate judgements, though sometimes preliminary test runs may be required. The panel size is around 25–30 people. They sample the product before large-scale consumer trials are conducted. They are adept in discriminating between samples and communicating their reactions in an effective way.

b. **Consumer panels:** These panels comprise those people who are the target group of the particular food product. They are untrained people chosen at random and surveyed largely by the questionnaire method. These panels should be made up of a large number of people in order to get a dependable and accurate result, including the opinion of a larger section of the population. A minimum group should be of 100 people.

Selection of the panel judges: Members of the panel should be carefully selected and trained to find out difference in specific quality characteristics between different stimuli. The requirements for ideal panel members are as follows:

1. Should be able to discriminate easily between the coded samples.

2. Should have good health, as a sick patient cannot judge the food correctly.
   Should not be a habitual consumer of chewing tobacco of pan (betel).

3. Mouth fresheners or peppermint should not be allowed before or after testing the samples.

4. The tester should neither be too hungry nor too full, as this may cloud the judgement.

5. Should be experienced in the particular field.

6. Should have high personal integrity. Should be able to evaluate objectively.

7. There should be intellectual curiosity.

8. Willingness to spend time for the sensory evaluation work is required.

9. Should be the ability to concentrate and learn.

10. Should have interest in sensory analysis of samples.

11. Should have ability to derive proper conclusion.

12. Should be available and willing to submit to periodic tests to get consistent results.

Candidates possessing these qualities are indexed with details of age, sex, specific likes and dislikes, availability.

o **Testing laboratory:** Testing laboratory consists of three separate units.

1. Reception room where the panel members meet the person in charge of the laboratory and get acquainted with the type of samples to be tested. This provides for people control.

2. The sample preparation room equipped for preparing and serving food. The storage area should be extra-large. The serving utensils should be plenty to avoid scarcity.

3. The test booths where the panel members carry out actual sensory evaluation of the samples. These should be located in close proximity to the other two areas.

The entire testing laboratory should be air-conditioned, free from noise and extraneous odours. Whenever samples with difference in colour are tested, lights should be used. The booths should be provided with adequate light, identical décor, a glass of water with clean drinking water, clean towels and platform for examination. Stainless steel, glass and dishes and cups and plain serving China should be used. Electric cooking equipment is preferred over gas equipment.

o **Preparation of samples:** Preparation should be done in ideal conditions and controlled atmosphere. The hygiene of the surroundings is very important for making correct judgements. Sample should be prepared from standardised ingredients in exact amounts. Careful labelling of sample and coding should be ensured. Samples for presentation must be from homogenous lot. Careful sampling of food is necessary for sensory evaluation. All samples should be at the same temperature, optimum level and kept constant during the test.

o **Techniques of smelling and tasting:** For odor tests of food products, a special technique is used to perceive aroma more clearly. Smelling is done with repeated short, rapid, sniffs. For taste tests, the sample is kept on the tongue for a while so that it reaches all parts of the tongue where the taste buds are located.

o **Testing time:** Testing should be done at a time when the panel members are fresh. The test time is generally between 10:00 am and 12:00 noon since this is the time when the senses are most active and the tester is neither too hungry nor full. Too many samples should not be given as they produce fatigue and lead to errors in results.

o **Designs of experiments:** Experimental error can be minimised through the use of techniques of randomizing. The experiment should be designed on the basis of the accuracy needed and the amount of sample available.

REASONS FOR TESTING FOOD QUALITY

o **To find out the consumer preference:** This helps the producer to discover which qualities of the product need to be developed and emphasised. The results are considered to represent the taste of the significant portion and are used to predict market outlook for a product.

o **Effect of variation in processing on quality:** Tests are done to investigate the influence of factors in production. Its purpose is to determine whether a given variation in processing has altered the quality of flavour of the products. It is also used to test the effect of storage and packaging.

o **To detect the presence of off-quality:** The panel members are usually trained to recognize and evaluate the standard flavour of food so that they can use their power of discrimination consistently.

EVALUATION CARD

The questionnaire or score card should be prepared carefully for each test. The card should be clearly typed or handwritten. It should be simple and use unambiguous terms and directions in the desired sequence of action. A scorecard in too much detail and cluster should be discouraged; too brief a form may fail to obtain some important information.

The design of the scorecards for sensory evaluation is challenging and difficult because the key characteristics of the product need to be evaluated on paper in a way that permits the judges to transmit their assessments of the samples accurately to the researcher. Different scorecards should be developed for each sample. A sheet for indicating rank order for a single characteristic should be attached.

OBJECTIVE EVALUATION

Food quality can also be successfully tested by objective methods. These techniques do not use the sensory perceptions as yardsticks for measurements but other characteristics of food. These methods use instruments or a standard method to evaluate food quality. These methods are called objective methods and the evaluation done is referred to as objective evaluation.

The various advantages of using these methods are:
1. The ease with which they can be reproduced. They are reproducible and yield the same results.
2. Human intervention is not involved and therefore fluctuations due to human sensitivity are not present.
3. The results obtained by these tests are accurate and minute differences obtained through sensory evaluation can be corrected.

4. They are less prone to errors as human involvement is to the minimum.
5. Fluctuations in measurement due to human emotional burdens or moods are avoided.
6. The measurements once made provide a permanent record and a database can be formed for future comparisons.
7. The external influences are avoided.
8. The effect of other factors besides the one being measured is avoided.
9. They give a feeling of confidence in the measurements.
10. The instruments used provide a sense of trust and belief in the results.

As often said "nothing is perfect", so are objective methods. They suffer from disadvantages too.

1. They consume more time than sensory evaluation.
2. Technical know-how and expertise is required to conduct these tests.
3. Specific equipment for the tests is required.
4. It is an expensive proposition.
5. Some characteristics of food like flavour and likes–dislikes of people cannot be measured by these methods.
6. In case of unavailability of skilled people or instruments the tests cannot be performed.
7. Untuned instruments can sometimes give wrong results.
8. Objective methods require calculations to be made which, if not done accurately, will yield wrong results.

To get very accurate results usually both sensory and objective methods are used. Objective evaluation supplements the data obtained through sensory evaluation.

Therefore, subjective and objective evaluations are generally advocated to be doubly sure of the results of evaluation of food quality.

Basic Directions

It is very important to know the correct way to perform the objective tests in order to obtain the desired results. If these instructions are taken into account, the results obtained will be accurate.

1. A maximum number of available tests should be conducted for evaluation of food. A composite and comparative analysis of these tests will help in arriving at the correct interpretation.
2. Advance planning should be done about the tests to be conducted and characteristics to be studied. Last moment worries and hurries can then be avoided.
3. Prior arrangements should be made before starting the tests. Availability of sample, equipment, skilled operator, etc., should be ensured.
4. There should be proper maintenance of the equipment to be used. It is necessary to check whether they are in perfect condition and give correct results.
5. A flowchart as to how the test has to proceed should be prepared in advance to avoid last moment confusions.
6. A dry run of the steps to be performed should be made to help in smooth running of the test.
7. The samples should be tested for constant properties before the onset of the experiment, e.g. while testing the tenderness of cakes, identical cakes should be taken as samples.

Objective Tests for Evaluation

There are various tests that have been devised in order to make deductions regarding the quality of food and analysis of the results obtained. The various tests have been grouped under the following heads:

1. Chemical methods
2. Physio-chemical methods
3. Microscopic examination
4. Physical methods

Chemical methods: Chemical estimation of food products is a useful criterion to judge its quality. Presence or absence of certain chemicals in foods makes them edible or non-edible. The chemical methods help in determining the quality of food and in analysis of quality. Iodine number, peroxide value, sugar tests, etc. are conducted to estimate the presence of chemicals in food and testing the quality. Food spoilage by peroxides in fats, adulterants in food, starch in milk, melanin yellow in turmeric, brick powder in chilli powder, etc. can be determined by chemical analysis.

Physio-chemical methods: These methods aid in determining the state of foods. Spoilt food can thus be found and discarded. Grading of food on the basis of some specific characteristic can be made and the best quality food selected.

Some physio-chemical tests are:

1. The measurement of pH of food products. Every food is best at an optimal pH. Any alteration in its pH will influence the quality of food. Therefore, by measuring the pH useful data regarding the quality of food can be found out. A pH meter is used for this purpose.

2. Sugar concentration is a vital property of jams, jellies, squashes, etc. and a useful index in preservation of foods. The measurement of sugar concentration is made by a refractometer. It is based on the principle that light is refracted as it passes through a sugar solution and indicates the degree of concentration. The refractometer is a calibrated instrument and values can be read directly from the instrument. Brix is the unit for sugar concentration and indicates percent sucrose in solution.

3. A quantitative analysis of sugar can be made by a polariscope.

Microscopic examinations: Sometimes, it is very difficult to assess the quality of food by observing external properties. The naked eye is not a reliable indicator of spoilage. In such cases, the help of the microscope is taken and it yields valuable data regarding food quality. These tests help in identifying and in some cases isolating the following:

1. Types of organisms present in fermented foods.
2. Starch cells in food.
3. Spoilage causing organisms in food.
4. Sugar crystals in food. Their shape and size.
5. Air cells in batters and foams.

6. Fungal contamination in cereals and fresh fruits and vegetables.
7. Spoilage microbes in milk and juices, etc.
8. Asymptomatic microbes in foods.

Physical methods: These tests measure the physical properties of food and on the basis of these make deductions regarding their quality.

1. **Weight:** A useful index for assessment, it indicates the degree of freshness of food, e.g. heavy eggs are better than light ones. Some fruits like apples and vegetables like brinjals are best when heavy. Bread and cakes are tastier and more acceptable when light and fluffy.

2. **Volume:** It is the space taken by an object. It is found by multiplying length, breadth and height of objects. Volume of liquids can be easily measured by measuring cups while volume of solid food is calculated by the displacement method. An easy way to find volume is by subtracting the volume of seeds held by a container with a product from that of volume of seeds without the product. Usually mustard seeds are used.

3. **Specific volume:** It is found by dividing the bulk volume by the weight of the product. It is very important to remember to take replicates when determining specific volume and then calculate the average with care as lot of experimental errors are likely. When measuring bulk volume for a porous product, it is important to seal off the porous areas with a substance like wax and then take the bulk volume, which is the increase in its volume. The displacement method with solvents is advocated for volume measurement.

   Specific volume = bulk volume / weight of product.

4. **Index to volume:** It is another indicator of food quality. It is found by measuring the volume of the most representative section of food, like the centre slice. The measurement is made by using the planimeter.

   The method used for the measurement is as follows:
   • Trace the detailed outline of the cross section of food. Use a sharp pencil or pen to make a clear outline.

- The product can also be traced by pressing it against a stamp pad and leaving an impression on paper as a blot.
- Another method is by using a planimeter. The complete outline, with protrusions, indentations and inundations is made carefully so that the final measurement recorded is the circumference/ perimeter of the slice.

5. **Specific gravity:** It is a very useful measure and is widely used in physical measurements. It is calculated by taking the density of water as a standard. It is the relative density of a substance when compared to water. It is calculated by dividing the weight of a given volume of substance with weight of the same volume of water. Lightness of products can be compared by this method and specifically for those which cannot be physically measured for volume.

   SG = weight of 'v' volume of substance / weight of 'v' volume of water

6. **Moisture:** Its presence in foods is an indicator of freshness and susceptibility to spoilage. It can be calculated by many methods some of which are mentioned below. As a general rule, the moisture content is expressed as percentage, i.e. moisture content = {(initial – final weight)/ initial weight × 100}.

   a. *Drying*: The sample of the substance is taken and weighed. The weight is noted. Thereafter the sample is dried in a muffle furnace till its weight becomes constant. All transfers of the sample should be carried out in desiccators. After this, the final weight is noted and the difference of the two values gives the moisture content.

   b. *Pressing*: This is another way of finding the moisture content of substances. It consists of noting the initial weight of a given quantity of the substance. Then for a specified period of time a known amount of pressure is applied on the object. This results in extraction of juices and fluids from the substance. The weight of the substance is again taken and the difference gives the amount of juice contained within the sample. This is used for finding out the juiciness in meat and flesh products.

c. *Titration:* Also known as Karl Fischer titration since it was discovered by a scientist with this name in 1990. It consists of homogenisation of the food to be analysed in a blender at 7500 rpm speed to release water from it. The released water is then titrated with the Karl Fischer reagent and the readings noted. The calculations are made by an inbuilt microprocessor. The machine is costly but very effective.

7. **Wet characteristics:** Refers to the property of foods to retain water and is also called wettability. It is generally used for baked products with air spaces in between. The baked product is weighed before the test and then placed in a dish of water for about 5 seconds. The product is then taken out and weighed again. The increase in the weight or weight gain gives the wettability indication. More moisture retention indicates good wettability.

8. **Cell structure:** It is an important criterion while judging the quality of baked products. It measures the uniformity, size, shape and thickness of the cell wall. This is done to get a three-dimensional view of the structure. A cut surface, when copied, gives an insight into the internal cell structure. Various methods have been developed to facilitate this study. Photocopying, photography and ink prints are useful in this direction. Size of the grain can be found out by making ink prints by using stamp pad. Alternatively, photocopies can be made. Another way is to take a photograph of the sample. This may not be representative of the true scale but it gives a general idea about the structure. A marked scale can be used to get the exact size.

9. **Colour:** It is a very important judgment criterion for foods. It is by far the easiest and commonest characteristic a consumer notices and observes in food. By studying the colour of food one can determine its quality. It is

generally seen that colour changes accompany spoilage of food. Even flavour component of food changes with colour. Many instruments and methods are used for the colour assessment.

a. *Dictionaries for colour*: Catalogues or databases for colour called dictionaries have been developed. Maerz and Paul's dictionary is commonly used. The dictionary has 56 charts. Seven main groups of hues are made in order as in spectra. Each group has 8 plates. This way colour matching from a standard can be made. There are two openings on the sheet. One is placed over the sample while the other is moved over colours and a match established.

b. *Disc colourimeter*: The discs used here have radial slits in them so that they can be fitted together. These are fitted such that each disc is visible in varying portions. These are then rotated at a speed of 2700 rpm such that a single hue develops without any flickering. For testing the sample it is placed adjacent to the disc and both viewed together.

c. *Tintometer*: It is another instrument used for colour determination. It is provided with standard permanent glass slides of red, yellow and blue colours. There are a range of slides from very light tints of colour designated with number 0 to dark coloured tints numbered 20. The sample is put such that half of viewing area is covered by it while the other half is rotated with the reflected light by those of slides. Matching of the sample is made.

d. *Coloured chips:* Food is also matched with chips, glass pieces of charts of standard colours. It is not a widely used method as it is not very practical and does not yield satisfactory results.

e. *Spectrometer:* It is an electrical instrument which works on the principle that different amounts of light are absorbed by different coloured substances. Light is made incident on the sample and the extent of absorbance or transmittance indicates the colour of the sample. Transmittance is 100% by a sample of distilled water.

Texture Evaluation

Texture is an important criterion for food acceptability. While some foods are preferred when grainy others are liked best when smooth. Therefore, a texture test is an acceptable test for food quality evaluation.

Various instruments have been developed which can measure the texture of liquids, semisolids and solids. The science of deformation and flow of water is termed rheology and the study is rheological studies. These studies are concerned with the measurement of mechanical properties of food. These studies gain importance since they are helpful in determining the flow properties of liquids as well as mechanical fate of solid foods when consumed and processed. These studies help in deciding which methods should be applied to foods while processing them.

Liquids and semisolids: These foods are those that do not have a rigid and definite shape. They either take the shape of the container, as liquids, or offer some resistance, like semisolids.

The property that determines this behaviour is viscosity or consistency. It is, therefore, very important to measure this quality in order to grade the products. The reasons for studying viscosity are:

1. It helps to decide whether the food is fully cooked or not, e.g. tomato ketchup by flow test.
2. Helps in finding out the quality of mayonnaise, ketchups, jams and jellies.
3. While cooking it helps to decide the duration and amount of heat to be applied.
4. During processing it helps to predict and check the final consistency of the product.
5. Viscosity measurements help in deciding the amount of thickening agents to be added in order to get a desired consistency.

Viscosity is nothing but the resistance offered by the liquid itself to the flow of liquids. It is the internal friction.

Instruments used for measurement of viscosity in liquids and semisolids are given as follows:

Instrument	Structure	Comments
Stormer viscometer	A container with a rotor which makes revolutions	Substance is put in the instrument and the time taken to make 100 revolutions helps to measure the consistency
Brookfield synchro-electric viscometer	A spindle immersed in a test substance	The resistance offered by the spindle to the sample is the measuring index
Bostwick consisto-meter	A channel (2'12") with 2" high sides. A centimetre scale on the channel flow and a triggered gate on one side	Based on the principle that length of flow of substance is proportional to consistency
Efflux-tube viscometer	Pressure applied on an instrument with an orifice or capillary	Time taken for the substance to pass through is measured
Adams consisto-meter	20 concentric circles (0.25" increased in radius) engraved upon a large metal disc. A steel cone which fits in the disc, its circumference being that of the innermost circle's radius. The cone can be lifted vertically.	The substance is filled in the cone and it is lifted for a while (30 seconds). The extent of flow is measured by recordings made at four equidistant points on the disc. The average is taken of the 4 values and consistency value calculated
Penetrometer	A plunger with a needle or cone which penetrates the sample by gravitational force for a decided period of time	The tenderness is measured by this instrument. More it penetrates or larger the distance it goes in the product, greater is the tenderness

Contd.

Instrument	Structure	Comments
Bloom gelometer	A cup, in which lead shots are added, is attached to a plunger and forces it in the sample	The lead shots are put in the cup one by one and this causes it to press the plunger to penetrate the sample. When sufficient shots have been put in the cup, the test is said to be completed and the number of shots required indicate the product's tenderness
Brabender farinograph	Mixer plates in a container in which sample is put	Used for testing plasticity of wheat dough for baked products. The instrument measures the force required to mix the dough and the effort put in by the mixer blades. There is an increase in force required in the beginning but then it decreases due to over mixing.

Percent sag: Another method to determine tenderness of a sample like gel or jelly. A probe is used to find the depth of the product while in its container. After this, the sample is demoulded to a flat plate. The percent sag is measured and calculated which indicates the tenderness of the product. Greater the value of percent sag, more tender the product.

$$\text{Percent sag} = \frac{\text{Depth in container} - \text{Depth in flat plate}}{\text{Depth in container}} \times 100$$

Solids: Solids are rigid and retain their shape irrespective of the container they are put in. Their external texture can be easily found out by touching but for the internal texture special methods are used. The principle behind the instruments is the resistance they offer to external force applied.

The terms used in this are:
1. *Compression*: When on pressing or squeezing food if it remains as one entity and is not fragmented.
2. *Cutting*: On application of force if the food divides into fragments it is said to be cut.

3. *Shearing*: This refers to partial separation of the food into parts which slide past one another but do not completely separate into pieces.
4. *Tensile strength*: When force is applied on food to pull it apart, it tests the tensile strength.

Instruments for determining properties in solids

Instrument	Structure	Comments
Magness-Taylor pressure test	A plunger attached with a spring which measures the compression force. The plunger is of variable diameter.	Works on the principle of compression. The plunger is pressed into the food to a set depth. The spring contracts and measures the compression force required to press the food.
Succulometer	An instrument that measures the volume of extracted liquid.	The food is compressed under controlled conditions of pressure for a given time period and the volume of the juice extracted is calculated.
Tenderometer	Measures the shearing as well as compression force.	The food is first compressed and then sheared to find its suitability, e.g. peas for preservation.
Fibrometer	An instrument that cuts the food into pieces.	This method helps to differentiate mature stocks from tender stocks, e.g. beans.
Shortometer	Platform with two parallel dull blades which hold the sample. The third blade presses down the sample when activated by a motor. This causes the sample to snap and break.	The force with which the sample snaps is an index of its tenderness. The less force it takes more tender the product is.
Christal texturo-meter	A series of rods that pierce the sample.	Generally used to test the hardness of meat. The principle used here is cutting.

Contd.

Instrument	Structure	Comments
Voldokevich bite tendero-meter	An instrument with sharp teeth in it. Utilises the principle of cutting and shearing.	The sample is put and the biting action, as of teeth, is imitated. This causes deformations in the food. The force required to do so indicates the texture of product, e.g. meat.
Kramer shear press	A multipurpose, multi-assembly instrument. The power unit is common but various assemblies can be made to conduct tests.	The sample is put in the press and tests conducted to determine the shearing force and action.
Compressimeter	Compression source to pressurise the sample till it deforms to a specific amount.	The sample is tested by measuring the force required to bring about a definite deformation. It measures the compressibi-lity and not shear strength.
Shear press	A pressure source.	It compresses, shears and extrudes the sample. Textural characteristics are measured by this press.
Warner Bratzler shear	Two parallel bars on opposite sides of a thin metal plate form the instrument.	Sample is placed on the metal plate and sheared by the bars. This measures the tenderness of meat.
Universal testing machine	A multipurpose machine.	It measures seven aspects of texture—cohesiveness, adhesiveness, hardness, springiness, gumminess, chewi- ness and fractura-bility.

Measurement of tensile strength: It is done to measure the tensile strength of products like chapatis. One end of the food is attached to a stand while the other remains in contact with a paper glass. Thereafter, the glass is filled with water till the weight is too much for the product to withstand. This breaks the product. If more water is required to do so, the product is strong and vice versa.

Measurement of extensibility: Food product to be measured is put in a grinder and greater the consumption of grinder more would be the toughness of the product. More power will be used. Extensibility is inversely related to tenderness. Therefore, more tender the product less extensible it will be.

Table 14.1: Comparison between sensory and objective tests

Sensory evaluation	Objective evaluation
Tests based on human senses.	Tests based on other methods.
Minimal use of any instrument, primarily human panel samples the food.	Primarily instruments and machines are used to make measurements and analysis.
Sense of smell, sight, taste and feel is employed.	Instruments of various kinds are used.
More prone to error.	Less prone to error.
Less accurate and consistent.	More accurate and consistent.
Are quick methods.	Are time consuming methods.
Are not exactly reproducible.	Are reproducible.
Internal problems of sickness and cold largely affect the result.	Human sickness has no effect on these results.
Results vary with each individual preference.	Instruments show same readings.
No permanent record can be made.	Databases can be maintained.
It is difficult to establish a standard.	Standards can be easily established.
Is inexpensive.	Is an expensive exercise.

QUALITY CONTROL

Foods are perishable by nature. Numerous changes take place in foods during processing and storage. It is well known that conditions used to process and store foods may adversely influence the quality attributes in foods. Up to storage for a certain period, one or more quality attributes of a food may reach an undesirable state. At that instant, the food is considered unsuitable for consumption and it is said to have reached the end of its shelf-life.

Shelf-life of a food product may be actually defined *as the time between the production and packaging of the product and the point at which it becomes unacceptable under defined environmental conditions.* Storage and distribution are necessary links in the food chain and hence considered as factors influencing the shelf-life.

Major causes of Food Deterioration

During storage and distribution, foods are exposed to a wide range of environmental conditions. Environmental factors such as temperature, humidity, oxygen and light can trigger several reaction mechanisms that may lead to food degradation. As a consequence of these mechanisms, food may be altered to such an extent that they are either rejected by the consumer or they may become harmful to the person consuming them. Chemical, physical and microbiological changes are the leading causes of the food deterioration.

Physical Changes

Physical changes are caused by mishandling of foods during harvesting, processing and distribution. These changes lead to reduced shelf-life of foods.

Crushing of dried snack during distribution seriously affects their quality.

Dried foods when kept in high humidity may pickup moisture and become soggy.

Chemical Changes

During the processing and storage of foods, several chemical changes occur that involve the internal food components and the external environmental factors. These changes may cause food deterioration and reduce the shelf-life. The most important chemical changes are associated with enzymatic action, oxidative reactions, particularly lipid oxidation that alters the flavour of many lipid containing foods and non-enzymatic browning that causes changes in appearance.

Fruits upon cutting tend to brown rapidly at room temperature due to the reaction of phenolase with the cell constituents that are released upon cutting of the tissue in presence of oxygen. Enzymes such as lipoxygenase, if not denatured during the blanching process, can influence food quality even at sub-freezing temperatures. In addition to the temperature, other environmental factors such as oxygen, water and pH induce deleterious changes in foods that are catalyzed by enzymes.

The presence of oil and fats containing unsaturated fatty acids is a prime reason for the development of rancidity in foods during storage as long as oxygen is available. Development of off-flavours which is markedly noticeable in rancid foods is the result of auto-oxidation of unsaturated fatty acids. The generation of free radicals during the autocatalytic process leads to other undesirable reactions, for example, loss of vitamins, alteration of colour and degradation of proteins. In addition to lipid oxidation, there are other chemical reactions that are induced by light such as loss of vitamins and browning of meats.

Non-enzymatic browning is a major cause of quality change and degradation of nutritional content in many foods. This type of browning reaction occurs due to the interaction between reducing sugars and amino acids. These reactions result in the loss of protein solubility, darkening of lightly coloured dry products and the development of bitter flavours. Environmental factors such as temperature, water activity and pH have an influence on non-enzymic browning.

Microbiological Changes

Microbes have the ability to multiply at high rates when favourable conditions are present. These factors are mainly temperature, pH, nutrition, etc. Prior to harvest, fruits and vegetables have generally good defence mechanisms against microbial attacks, however, after separation from the plant they can easily succumb to microbial proliferation. Similarly, meat upon slaughter is unable to resist rapidly growing microbes.

Microbial growth in foods results in food spoilage with the development of undesirable sensory characteristics and in certain cases the food may become unsafe for consumption.

The pathogenicity of certain micro-organisms is a major safety concern in processing and handling of foods upon ingestion. Micro-organisms such as *Salmonella* species and *Escherichia coli* strains cause infection while others such as *Aspergillus flavus*, *Clostridium botulinum* and *Staphylococcus aureus* produce chemicals in foods that are toxic to humans.

Evaluation of the Food Quality

A common practice employed to evaluate the shelf-life of a given food product is to determine changes in selected quality characteristics over a period of time. One may consider quality of a food as a gross measure of the food deterioration occurring in food item. However, it should be recognized that the term quality is meant to encompass several quality attributes or characteristics. From a consumer's standpoint, the sensory expectations derived from the presence (or absence) of desirable (or undesirable) characteristics of a given food determine the quality of a product. Therefore, a food product noted for its high quality has more of the desirable characteristics.

Empirical or analytical techniques may be used to quantify the quality attributes of food. For example, enumeration of microbes or determination of chemical components of a product are *analytical techniques*, whereas the human subjects to monitor changes in the magnitudes of quality characteristics constitute *empirical techniques*.

Procedures for Determination and Monitoring of Shelf-Life

Direct shelf-life determination requires batches of samples to be taken at significant stages in the development or modification of the product. These samples should be examined during storage, usually under controlled environmental conditions, until their quality becomes unacceptable. *The time when this occurs is the maximum product shelf-life*, and therefore the determination necessarily requires at least this time to complete.

Significant sampling stages within the programme of shelf-life evaluation include:

1. The successful *experimental kitchen or pilot plant batch*. At this stage, it is possible to investigate formulation, process or packaging changes to improve the shelf-

life without the costs of factory time and material quantities.

2. The successful *full scale factory batch*. This is the most important sampling stage. This will provide the data for the setting of shelf-life and specification standards.

3. The first *continuous production trial*. Examination of products should confirm the data from earlier samplings.

As part of an on-going surveillance system, samples should be taken at suitable intervals for storage trial. The sampling interval should typically be 20% of the shelf-life which will provide samples of 6 different ages from fresh to full shelf-life. For long life products, more frequent intervals may be useful to detect any changes in storage performance (e.g. every two months for a two-year shelf-life).

Shelf-life samples should be subjected to conditions effectively simulating the normal storage and distribution conditions the food is likely to encounter. Shelf-life examination is done by employing appropriate methods of sensory evaluation, chemical analysis and microbiological analysis.

In *sensory evaluation*, appearance, smell, texture and flavour being the main attributes to assess. Such assessments are frequently not exact as there may not be a suitable control sample with which to compare the stored samples, this being particularly so for new products. However, under appropriate test conditions, using control sample, it is possible to get a fair idea about the quality and acceptability of stored products.

Quantitative measurements, for example of colour, texture, viscosity and amount of water or oil separation should be included if they either closely relate to the sensory quality or can be used as reliable indicators of quality deterioration.

In addition to subjective assessments, other tests may be necessary. These may include tin content of products in unlacquered cans, vitamin content where a claim is made. *Microbiological examination* of fresh and stored products is highly essential in order to determine whether they are safe for human consumption.

Chapter 15

Extrepreneurship

Entrepreneurship has traditionally been defined as the process of designing, launching and running a new business, which typically begins as a small business, such as a start-up company, offering a product, process or service for sale or hire, and the people who do so are called 'entrepreneurs'.

It has been defined as the "capacity and willingness to develop, organize, and manage a business venture along with any of its risks in order to make a profit."

While definitions of entrepreneurship typically focus on the launching and running of businesses, due to the high risks involved in launching a start-up, a significant proportion of businesses have to close, due to a "lack of funding, bad business decisions, an economic crisis, or a combination of all of these" or due to lack of market demand.

In the 2000s, the definition of "entrepreneurship" has been expanded to explain how and why some individuals (or teams) identify opportunities, evaluate them as viable, and then decide to exploit them, whereas others do not, and, in turn, how entrepreneurs use these opportunities to develop new products or services, launch new firms or even new industries and create wealth.

Recent advances stress the fundamentally uncertain nature of the entrepreneurial process, because although opportunities exist their existence cannot be discovered or identified prior to their actualization into profits.

What appears as a real opportunity ex-ante might actually be a non-opportunity or one that cannot be actualized by entrepreneurs lacking the necessary business skills, financial capital or social capital.

Entrepreneurship is the process by which an individual (or team) identifies a business opportunity and acquires and deploys the necessary resources required for its exploitation. The exploitation of entrepreneurial opportunities may include actions such as developing a business plan, hiring the human resources, acquiring financial and material resources, providing leadership, and being responsible for the venture's success or failure.

Traditionally, an **entrepreneur** has been defined as "a person who starts, organizes and manages any enterprise, especially a business, usually with considerable initiative and risk".

"Rather than working as an employee, an entrepreneur runs a small business and assumes all the risk and reward of a given business venture, idea, or good or service offered for sale. The entrepreneur is commonly seen as a business leader and innovator of new ideas and business processes."

Entrepreneurs tend to be good at perceiving new business opportunities and they often exhibit positive biases in their perception (i.e. a bias towards finding new possibilities and seeing unmet market needs) and a pro-risk-taking attitude that makes them more likely to exploit the opportunity.

An entrepreneur is typically in control of a commercial undertaking, directing the factors of production—the human, financial and material resources—that are required to exploit a business opportunity. They act as the manager and oversee the launch and growth of an enterprise.

"Entrepreneurial spirit is characterized by innovation and risk-taking." While entrepreneurship is often associated with new, small, for-profit start-ups, entrepreneurial behaviour can be seen in small-, medium- and large-sized firms, new and established firms and in for-profit and not-for-profit organizations, including voluntary sector groups, charitable organizations and government.

The Entrepreneurship Ecosystem

Entrepreneurship typically operates within an entrepreneurship ecosystem which often includes government programs and services that promote entrepreneurship and support entrepreneurs and start-ups; non-governmental organizations such as small business associations and organizations that offer advice and mentoring to entrepreneurs (e.g. through entrepreneurship centres or websites); small business advocacy organizations that lobby the government for increased support for entrepreneurship programs and more small business-friendly laws and regulations; entrepreneurship resources and facilities (e.g. business incubators and seed accelerators); entrepreneurship education and training programs offered by schools, colleges and universities; and financing (e.g. bank loans, venture capital financing, angel investing, and government and private foundation grants).

Selection of an Ideal Plant Location for the Production

Plant location refers to the choice of region and the selection of a particular site for setting up a business or factory.

But the choice is made only after considering cost and benefits of different alternative sites. It is a strategic decision that cannot be changed once taken. If at all changed only at considerable loss, the location should be selected as per its own requirements and circumstances. Each individual plant is a case in itself. Businessman should try to make an attempt for optimum or ideal location.

Earlier this year, *Food Processing* magazine's Kevin T. Higgins claimed that being a food manufacturer was the next best thing to being a king.

This is because most new plant projects and renovations are now supported by generous, multimillion public subsidies.

A sound business plan should be the foundation of the site-selection process, detailing facts including:

o The goods the plant will produce
o The quantity of goods the plant will produce
o Five years of production planning
o Future growth expectations

Defining these facts are key because they impact various aspects of a site ranging from size requirements to energy availability.

The five important factors to consider while selecting plant location are:

1. *Local geography*: A site may require additional land depending on local conditions such as topography, drainage or governmental regulation.

2. *Daily operations*: How much traffic will be within the plant and how often? Consider activities such as raw ingredient delivery, finished product distribution and the facility's internal material movement. "Will the plant's operations require rail access?" or "Is interstate accessibility high priority?"

3. *Utility and water costs*: Determine the square footage one will require, and use a utility matrix to estimate its utility requirements. If your organization has built similar facilities, project these items based on existing data. Utility and water costs go hand-in-hand with the operation costs.

   When selecting a site, ensure an adequate energy supply source is readily available. Review the water and wastewater requirements. Water consumption costs are heavily influenced by wastewater pre-treatment costs or additional wastewater surcharges. Having estimates of these costs can help qualify the prospective sites.

4. *Distance*: Identify the location of the customers and distribution centres. Where are the raw materials located? Pinpointing these can help determine the ideal location of the new plant.

   Keep transportation costs in mind, as well. Understand that costs vary between inbound material and outbound product. One should strive to minimize the total transportation cost over simply minimizing the mileage between customers, materials and the plant.

5. *Environmental issues*: This is a crucial consideration for food processing plants. Examine nearby industrial sites. Do they emit dust? Is there air pollution? What

about odors? Noxious or toxic elements? If the site is contaminated, it is absolutely not suitable for a food processing facility. These issues are typically covered in a phase I environmental assessment, usually conducted by a consultant.

IDEAL PLANT LOCATION

An ideal location is one where the cost of the product is kept to minimum, with a large market share, the least risk and the maximum social gain. It is the place of maximum net advantage or which gives lowest unit cost of production and distribution. For achieving this objective, small-scale entrepreneur can make use of locational analysis for this purpose.

Locational analysis is a dynamic process where entrepreneur analyses and compares the appropriateness or otherwise of alternative sites with the aim of selecting the best site for a given enterprise. It consists of the following:

a. *Demographic analysis*: It involves study of population in the area in terms of total population (in number), age composition, per capita income, educational level, occupational structure, etc.

b. *Trade area analysis*: It is an analysis of the geographic area that provides continued clientele to the firm. He would also see the feasibility of accessing the trade area from alternative sites.

c. *Competitive analysis*: It helps to judge the nature, location, size and quality of competition in a given trade area.

d. *Traffic analysis*: To have a rough idea about the number of potential customers passing by the proposed site during the working hours of the shop, the traffic analysis aims at judging the alternative sites in terms of pedestrian and vehicular traffic passing a site.

e. *Site economics*: Alternative sites are evaluated in terms of establishment costs and operational costs under this. Costs of establishment is basically cost incurred for permanent physical facilities but operational costs are incurred for running business on day-to-day basis, they are also called running costs.

The important considerations for selecting a suitable location are given as follows:

a. Natural or climatic conditions.

b. Availability and nearness to the sources of raw material.

c. Transport costs—in obtaining raw material and also distribution or marketing finished products to the ultimate users.

d. *Access to market*: Small businesses in retail or wholesale or services should be located within the vicinity of densely populated areas.

e. Availability of Infrastructural facilities such as developed industrial sheds or sites, link roads, nearness to railway stations, airports or sea ports, availability of electricity, water, public utilities, civil amenities and means of communication are important, especially for small-scale businesses.

f. Availability of skilled and non-skilled labour and technically qualified and trained managers.

g. Banking and financial institutions are located nearby.

h. *Locations with links*: To develop industrial areas or business centers result in savings and cost reductions in transport overheads, miscellaneous expenses.

i. Strategic considerations of safety and security should be given due importance.

j. *Government influences*: Both positive and negative incentives to motivate an entrepreneur to choose a particular location are made available. Positive includes cheap overhead facilities like electricity, banking transport, tax relief, subsidies and liberalization. Negative incentives are in form of restrictions for setting up industries in urban areas for reasons of pollution control and decentralization of industries.

k. Residence of small business entrepreneurs want to set up nearby their homelands

Plant Layout

Plant layout refers to the arrangement of physical facilities such as machinery, equipment, furniture, etc. within the factory building in such a manner so as to have the quickest flow of material at the lowest cost and with the least amount of handling in processing the product from the receipt of material to the shipment of the finished product.

According to Riggs, "the overall objective of plant layout is to design a physical arrangement that most economically meets the required output—quantity and quality."

According to JL Zundi, "Plant layout ideally involves allocation of space and arrangement of equipment in such a manner that overall operating costs are minimized.

Plant layout is an important decision as it represents long-term commitment. An ideal plant layout should provide the optimum relationship among output, floor area and manufacturing process. It facilitates the production process, minimizes material handling, time and cost, and allows flexibility of operations, easy production flow, makes economic use of the building, promotes effective utilization of manpower, and provides for employee's convenience, safety, comfort at work, maximum exposure to natural light and ventilation. It is also important because it affects the flow of material and processes, labor efficiency, supervision and control, use of space and expansion possibilities, etc.

An efficient plant layout is one that can be instrumental in achieving the following objectives:

a. Proper and efficient utilization of available floor space
b. To ensure that work proceeds from one point to another point without any delay
c. Provide enough production capacity.
d. Reduce material handling costs
e. Reduce hazards to personnel
f. Utilise labor efficiently
g. Increase employee morale
h. Reduce accidents
i. Provide for volume and product flexibility
j. Provide ease of supervision and control
k. Provide for employee safety and health
1. Allow ease of maintenance

m. Allow high machine or equipment utilization

n. Improve productivity

Layout Types

From the point of view of plant layout, one can classify small business or unit into three categories:

1. *Manufacturing units*

In case of manufacturing unit, plant layout may be of four types:

a. Product or line layout

b. Process or functional layout

c. Fixed position or location layout

d. Combined or group layout

a. Product or line layout: Under this, machines and equipment are arranged in one line depending upon the sequence of operations required for the product. The materials move from one workstation to another sequentially without any backtracking or deviation. Under this, machines are grouped in one sequence. Therefore materials are fed into the first machine and finished goods travel automatically from machine to machine, the output of one machine becoming input of the next, e.g. in a paper mill, bamboos are fed into the machine at one end and paper comes out at the other end. The raw material moves very fast from one workstation to other stations with a minimum work in progress storage and material handling.

The grouping of machines should be done keeping in mind the following general principles.

○ All the machine tools or other items of equipment must be placed at the point demanded by the sequence of operations

○ There should no points where one line crossed another line.

○ Materials may be fed where they are required for assembly but not necessarily at one point.

    o All the operations including assembly, testing packing must be included in the line

(b) Process layout

In this type of layout machines of a similar type are arranged together at one place.

For example, machines performing drilling operations are arranged in the drilling department, machines performing casting operations be grouped in the casting department. Therefore, the machines are installed in the plants, which follow the process layout.

Hence, such layouts typically have drilling department, milling department, welding department, heating department and painting department, etc. The process or functional layout is followed from historical period. It evolved from the handicraft method of production. The work has to be allocated to each department in such a way that no machines are chosen to do as many different job as possible, i.e. the emphasis is on general purpose machine.

The work, which has to be done, is allocated to the machines according to loading schedules with the object of ensuring that each machine is fully loaded.

The grouping of machines according to the process has to be done keeping in mind the following principles

    o The distance between departments should be as short as possible for avoiding long distance movement of materials

    o The departments should be in sequence of operations

    o The arrangement should be convenient for inspection and supervision

(c) Fixed Position or Location Layout

In this type of layout, the major product being produced is fixed at one location. Equipment labor and components are moved to that location. All facilities are brought and arranged around one work centre. This type of layout is not relevant for small-scale entrepreneur.

(d) Combined layout

Certain manufacturing units may require all three processes, namely intermittent process (job shops), the continuous process (mass production shops) and the representative process combined process [i.e. miscellaneous shops].

In most of industries, only a product layout or process layout or fixed location layout does not exist. Thus, in manufacturing concerns where several products are produced in repeated numbers with no likelihood of continuous production, combined layout is followed. Generally, a combination of the product and process layout or other combinations are found, in practice, e.g. for industries involving the fabrication of parts and assembly, fabrication tends to employ the process layout, while the assembly areas often employ the product layout.

Advantages, Disadvantages and Suitability of Plant Layouts

Plant layout	Advantages	Disadvantages	Suitability
Product or line layout	Low cost of material handling, due to straight and short route and absence of backtracking	High initial capital investment in special purpose machine	Mass production of standardized products
	Smooth and uninterrupted operations	Heavy overhead charges	Simple and repetitive manufacturing process
	Continuous flow of work	Breakdown of one machine will hamper the whole production process	Operation time for different process is more or less equal
	Lesser investment in inventory and work in progress	Lesser flexibility as specially laid out for particular product.	Reasonably stable demand for the product

Contd.

Plant layout	Advantages	Disadvantages	Suitability
	Optimum use of floor space		Continuous supply of materials
	Shorter processing time or quicker output		
	Less congestion of work in the process		
	Simple and effective inspection of work and simplified production control		
	Lower cost of manufacturing per unit		
Process Layout	Lower initial capital investment in machines and equipment.	Material handling costs are high due to backtracking	Products are not standardized
	There is high degree of machine utilization, as a machine is not blocked for a single product.	More skilled labor is required resulting in higher cost.	Quantity produced is small
	The overhead costs are relatively low.	Time gap or lag in production is higher	There are frequent changes in design and style of product
	Change in output design and volume can be more easily adapted to the output of variety of products	Work in progress inventory is high needing greater storage space	Job shop type of work is done

Contd.

Plant layout	Advantages	Disadvantages	Suitability
	Breakdown of one machine does not result in complete work stoppage	More frequent inspection is needed which results in costly supervision	Machines are very expensive
	Supervision can be more effective and specialized		
	There is a greater flexibility of scope for expansion.		
Fixed Position Layout	It saves time and cost involved on the movement of work from one workstation to another.	Production period being very long, capital investment is very heavy	Manufacture of bulky and heavy products such as locomotives, ships, boilers, generators, wagon building, aircraft manufacturing, etc.
	The layout is flexible as change in job design and operation sequence can be easily incorporated.	Very large space is required for storage of material and equipment near the product.	Construction of building, flyovers, dams.
	It is more economical when several orders in different stages of progress are being executed simultaneously. Adjustments can be made to meet shortage of materials or absence of workers by changing the sequence of operations.	As several operations are often carried out simultaneously, there is possibility of confusion and conflicts among different work groups.	Hospital, the medicines, doctors and nurses are taken to the patient (product).

2. Traders

When two outlets carry almost same merchandise, customers usually buy in the one that is more appealing to them. Thus, customers are attracted and kept by good layout, i.e. good lighting, attractive colours, good ventilation, air conditioning, modern design and arrangement and even music. All of these things mean customer convenience, customer appeal and greater business volume.

The customer is always impressed by service, efficiency and quality. Hence, the layout is essential for handling merchandise, which is arranged as per the space available and the type and magnitude of goods to be sold keeping in mind the convenience of customers.

There are three kinds of layouts in retail operations today.
o Self-service or modified self-service layout
o Full service layout
o Special layouts

3. Services Centres and Establishment

Services establishments such as motels, hotels, restaurants, must give due attention to client convenience, quality of service, efficiency in delivering services and pleasing office ambience. In today's environment, the clients look for ease in approaching different departments of a service organization and hence the layout should be designed in a fashion, which allows clients quick and convenient access to the facilities offered by a service establishment.

FACTORS DICTATING PLANT LAYOUT

a. **Factory building:** The nature and size of the building determines the floor space available for layout. While designing the special requirements, e.g. air conditioning, dust control, humidity control, etc. must be kept in mind.

b. **Nature of product:** Product layout is suitable for uniform products, whereas process layout is more appropriate for custom-made products.

c. **Production process:** In assembly line industries, product layout is better. In job order or intermittent manufacturing on the other hand, process layout is desirable.

d. **Type of machinery:** General purpose machines are often arranged as per process layout while special purpose machines are arranged according to product layout.

e. **Repairs and maintenance:** Machines should be so arranged that adequate space is available between them for movement of equipment and people required for repairing the machines.

f. **Human needs:** Adequate arrangement should be made for cloakroom, washroom, lockers, drinking water, toilets and other employee facilities, proper provision should be made for disposal of effluents, if any.

g. **Plant environment:** Heat, light, noise, ventilation and other aspects should be duly considered, e.g. paint shops and plating section should be located in another hall so that dangerous fumes can be removed through proper ventilation, etc. Adequate safety arrangement should also be made.

Thus, the layout should be conducive to health and safety of employees. It should ensure free and efficient flow of men and materials. Future expansion and diversification may also be considered while planning factory layout.

FINANCING OF THE BUSINESS VENTURE

Many companies face the challenge of raising the necessary capital to start a business, finance day-to-day operations or fuel the development and growth of their enterprise. It is not uncommon for many entrepreneurs start their businesses with their own funds or with help from family and friends. However, growing the business may necessitate purchasing inventory, upgrading technology or acquiring assets, for which the cash flow of the business alone may prove inadequate. Typically, these enterprises turn to outside funding to fuel their growth.

According to Don Hofstrand, retired extension value added agriculture specialist (Iowa State University), financing is needed to start a business and ramp it up to profitability. There are several sources to consider when looking for start-up financing. But first you need to consider how much money you need and when you will need it.

The financial needs of a business will vary according to the type and size of the business. For example, processing businesses are usually capital intensive, requiring large amounts of capital. Retail businesses usually require less capital.

Debt and equity are the two major sources of financing. Government grants to finance certain aspects of a business may be an option. Also, incentives may be available to locate in certain communities and/or encourage activities in particular industries.

Types

Business finance is available in multiple forms, including term loans, short-term loans, equipment financing, factoring, capital from angel investors and credit card loans.

Term loans are very popular loans for many companies in need of money for expansion, acquisition, or working capital and can also be used for refinance. This type of business finance loan is usually repaid over a period of time based on the useful life cycle of the asset class being purchased.

Factoring consists of selling the account receivables or invoices.

Receivables are the payments owed by customers to whom one has extended credit. The factor buys the receivables and gives a percentage of their total value, such as 80 percent. This is called an advance.

When the customers pay the factor, the remaining amount is paid minus the cost of factoring. Usually, the fee for this service is 3 to 5 percent of the total amount of the receivables.

Angel investors are usually private sources who are willing to invest their capital in start-ups or companies with potentially groundbreaking products or services.

Businesses can use credit cards to their advantage as another form of financing. Many of the cards have no annual fee, offer generous rewards programs and hefty credit limits. Besides using credit cards to track and manage their business expenses, many entrepreneurs use them to help ease their way through when experiencing a cash-flow crunch.

Equity Financing

Equity financing means exchanging a portion of the ownership of the business for a financial investment in the business. The ownership stake resulting from an equity investment allows the investor to share in the company's profits. Equity involves a permanent investment in a company and is not repaid by the company at a later date.

The investment should be properly defined in a formally created business entity. An equity stake in a company can be in the form of membership units, as in the case of a limited liability company or in the form of common or preferred stock as in a corporation. Companies may establish different classes of stock to control voting rights among shareholders. Similarly, companies may use different types of preferred stock. For example, common stockholders can vote while preferred stockholders generally cannot. But common stockholders are last in line for the company's assets in case of default or bankruptcy. Preferred stockholders receive a predetermined dividend before common stockholders receive a dividend.

Personal Savings

The first place to look for money is your own savings or equity. Personal resources can include profit-sharing or early retirement funds, real estate equity loans, or cash value insurance policies.

Life insurance policies: A standard feature of many life insurance policies is the owner's ability to borrow against the cash value of the policy. This does not include term insurance because it has no cash value. The money can be used for business needs. It takes about two years for a policy to accumulate sufficient cash value for borrowing. You may borrow most of the cash value of the policy. The loan will reduce the face value of the policy and, in the case of death, the loan has to be repaid before the beneficiaries of the policy receive any payment.

Home equity loans: A home equity loan is a loan backed by the value of the equity in your home. If your home is paid for, it can be used to generate funds from the entire value of your home. If your home has an existing mortgage, it can provide funds on the difference between the value of the house and the unpaid mortgage amount. Some home equity loans are set up as a revolving credit line from which you can draw the amount needed at any time. The interest on a home equity loan is tax deductible.

Friends and Relatives

Founders of a start-up business may look to private financing sources such as parents or friends. It may be in the form of equity financing in which the friend or relative receives an ownership interest in the business. However, these investments should be made with the same formality that would be used with outside investors.

Venture Capital

Venture capital refers to financing that comes from companies or individuals in the business of investing in young, privately held businesses. They provide capital to young businesses in exchange for an ownership share of the business. Venture capital firms usually do not want to participate in the initial financing of a business unless the company has management with a proven track record. Generally, they prefer to invest in companies that have received significant equity investments from the founders and are already profitable.

They also prefer businesses that have a competitive advantage or a strong value proposition in the form of a patent, a proven demand for the product, or a very special (and protectable) idea. Venture capital investors often take a hands-on approach to their investments, requiring representation on the board of directors and sometimes the hiring of managers. Venture capital investors can provide valuable guidance and business advice. However, they are looking for substantial returns on their investments and their objectives may be at cross purposes with those of the founders. They are often focused on short-term gain.

Venture capital firms are usually focused on creating an investment portfolio of businesses with high-growth potential resulting in high rates of returns. These businesses are often high-risk investments. They may look for annual returns of 25 to 30 percent on their overall investment portfolio.

Because these are usually high-risk business investments, they want investments with expected returns of 50 percent or more. Assuming that some business investments will return 50 percent or more while others will fail, it is hoped that the overall portfolio will return 25 to 30 percent.

More specifically, many venture capitalists subscribe to the 2-6-2 rule of thumb. This means that typically two investments will yield high returns, six will yield moderate returns (or just return their original investment), and two will fail.

Angel Investors

Angel investors are individuals and businesses that are interested in helping small businesses survive and grow. So their objective may be more than just focusing on economic returns. Although angel investors often have somewhat of a mission focus, they are still interested in profitability and security for their investment. So they may still make many of the same demands as a venture capitalist.

Angel investors may be interested in the economic development of a specific geographic area in which they are located. Angel investors may focus on earlier stage financing and smaller financing amounts than venture capitalists.

Government Grants

Federal and state governments often have financial assistance in the form of grants and/or tax credits for start-up or expanding businesses.

Equity Offerings

In this situation, the business sells stock directly to the public. Depending on the circumstances, equity offerings can raise substantial amounts of funds. The structure of the offering can take many forms and requires careful oversight by the company's legal representative.

Initial Public Offerings

Initial Public Offerings (IPOs) are used when companies have profitable operations, management stability, and strong demand for their products or services. This generally does not happen until companies have been in business for several years. To get to this point, they usually will raise funds privately one or more times.

Warrants

Warrants are a special type of instrument used for long-term financing. They are useful for start-up companies to encourage investment by minimizing downside risk while providing upside potential. For example, warrants can be issued to management in a start-up company as part of the reimbursement package.

A warrant is a security that grants the owner of the warrant the right to buy stock in the issuing company at a pre-determined (exercise) price at a future date (before a specified expiration date). Its value is the relationship of the market price of the stock to the purchase price (warrant price) of the stock. If the market price of the stock rises above the warrant price, the holder can exercise the warrant. This involves purchasing the stock at the warrant price. So, in this situation, the warrant provides the opportunity to purchase the stock at a price below current market price.

If the current market price of the stock is below the warrant price, the warrant is worthless because exercising the warrant would be the same as buying the stock at a price higher than the current market price. So, the warrant is left to expire. Generally warrants contain a specific date at which they expire if not exercised by that date.

Debt Financing

Debt financing involves borrowing funds from creditors with the stipulation of repaying the borrowed funds plus interest at a specified future time. For the creditors (those lending the funds to the business), the reward for providing the debt financing is the interest on the amount lent to the borrower.

Debt financing may be secured or unsecured. Secured debt has collateral (a valuable asset which the lender can attach to satisfy the loan in case of default by the borrower). Conversely, unsecured debt does not have collateral and places the lender in a less secure position relative to repayment in case of default.

Debt financing (loans) may be short term or long term in their repayment schedules. Generally, short-term debt is used to finance current activities such as operations while long-term debt is used to finance assets such as buildings and equipment.

Friends and Relatives

Founders of start-up businesses may look to private sources such as family and friends when starting a business. This may be in the form of debt capital at a low interest rate. However, if you borrow from relatives or friends, it should be done with the same formality as if it were borrowed from a commercial lender. This means creating and executing a formal loan document that includes the amount borrowed, the interest rate, specific repayment terms (based on the projected cash flow of the start-up business), and collateral in case of default.

Banks and Other Commercial Lenders

Banks and other commercial lenders are popular sources of business financing. Most lenders require a solid business plan, positive track record, and plenty of collateral. These are usually hard to come by for a start-up business. Once the business is underway and profit and loss statements, cash flows budgets, and net worth statements are provided, the company may be able to borrow additional funds.

Commercial Finance Companies

Commercial finance companies may be considered when the business is unable to secure financing from other commercial sources. These companies may be more willing to rely on the quality of the collateral to repay the loan than the track record or profit projections of your business. If the business does not have substantial personal assets or collateral, a commercial finance company may not be the best place to secure financing. Also, the cost of finance company money is usually higher than other commercial lenders.

Government Programs

Federal, state, and local governments have programs designed to assist the financing of new ventures and small businesses. The assistance is often in the form of a government guarantee of the repayment of a loan from a conventional lender. The guarantee provides the lender repayment assurance for a loan to a business that may have limited assets available for collateral. The best known sources are the Small Business Administration and the USDA Rural Development programs.

Bonds

Bonds may be used to raise financing for a specific activity. They are a special type of debt financing because the debt instrument is issued by the company. Bonds are different from other debt financing instruments because the company specifies the interest rate and when the company will pay back the principal (maturity date). Also, the company does not have to make any payments on the principal (and may not make any interest payments) until the specified maturity date. The price paid for the bond at the time it is issued is called its face value.

When a company issues a bond it guarantees to pay back the principal (face value) plus interest. From a financing perspective, issuing a bond offers the company the opportunity to access financing without having to pay it back until it has successfully applied the funds. The risk for the investor is that the company will default or go bankrupt before the maturity date. However, because bonds are a debt instrument, they are ahead of equity holders for company assets.

Lease

A lease is a method of obtaining the use of assets for the business without using debt or equity financing. It is a legal agreement between two parties that specifies the terms and conditions for the rental use of a tangible resource such as a building and equipment. Lease payments are often due annually. The agreement is usually between the company and a leasing or financing organization and not directly between the company and the organization providing the assets. When the lease ends, the asset is returned to the owner, the lease is renewed, or the asset is purchased.

A lease may have an advantage because it does not tie up funds from purchasing an asset. It is often compared to purchasing an asset with debt financing where the debt repayment is spread over a period of years. However, lease payments often come at the beginning of the year where debt payments come at the end of the year. So, the business may have more time to generate funds for debt payments, although a down payment is usually required at the beginning of the loan period.

FINANCING OF THE BUSINESS VENTURE SPECIFIC TO THE INDIAN CONTEXT

Entrepreneurs and business owners know that a business runs on working capital. Capital required to take the business to the next level is available through a business loan. As a Small and Medium Enterprise (SME) owner in India, one can take advantage of various options through the different types of the business loans in India targeted specifically to the particular industry and type of business.

Types of Business Loans in India

SBA Business Loan: The SBA loan is a loan offered by banks and NBFCs who have been guaranteed by the SBA (Small Business Association). The lenders are usually in the private sector and help finance small businesses. The SBA does not have the funds to finance long-term businesses, thus it works with these lenders to help give fixed rate financing for lands and buildings and other equipment that may be required.

Working Capital Loans: Working capital loans are loans taken in order to finance daily activities in order to keep the business running. These loans are of two types—secured and unsecured working capital loans.

    a. *The secured working capital loans* are business loans given by the lender against assets such as equipment, buildings or accounts, as security or collateral. The lender assesses and then decides the amount that can be given against these assets. Sometimes personal assets such as homes or shares may also be required to be kept as collateral.

b. *The unsecured working capital loans* are loans that are given without any assets or collateral for security. These loans are as such difficult to obtain as there is no security against these loans.

Commercial real estate loans are loans given on commercial real estates. Commercial real estates are properties or real estates used only for business purposes such as apartments, office complexes or retail centers. The commercial real estate loans are generally procured for the development, acquisition and construction of these lands or properties. These loans can generally be secured from banks or private sector lenders.

Start-up loans as the name suggests are loans that are usually availed to start a new business venture. In order to be eligible for this loan you should have your business ideas on paper and they should be conceivable. This loan can be usually procured from banks and private sector lenders and sometimes requires personal assets to be kept as collateral.

Professional loans are loans offered to professionals such as dentists or lawyers or doctors or other such professionals who are starting their own practices or firms. These include CPAs also.

Hard Money Equity Loans, Multi-Family Real Estate Loans and Business Acquisition Loans are offered by the banks and other lenders according to the need and qualification of the consumer.

Hard money equity loans are loans that are given on specific real property. These credits are usually given by companies or private investors. They are usually given for projects that are of short duration, such as for a few years or a few months only. Due to this the rate of interest on the loans is higher and they have a high-risk factor.

Business acquisition loans are loans taken from banks or other private sector lenders or investors in order to acquire a business that already exists. We can call it a takeover or expansion by a businessman who currently owns a running business and is looking to expand it.

Now we know that the various business loans available in India, we can dive deeper into the types of small business loans as well. Business can be broadly categorized into two types, namely small scale and large scale businesses. Similarly, there is a subcategory to business loans that offer loans to small scale businesses as well and are called small business loans.

Types of small business loans:
o Term loan
o Loan against property
o Gold loan
o Loan against shares and mutual funds
o Cash credit facility
o Letter of credit facility

Apart from loans, there are other financing options also available to finance the business or for its expansion or other needs that need considerable capital to create or buy.

Types of business financing options
o Bootstrapping
o Family and friends
o Crowdfunding
o Equity financing
o Debt financing
o Angel investors
o Leases
o Government programs

These are a few alternate forms of financing that can be opted for in case you are not eligible for business loans or do not wish to apply for business loans from banks or other lenders. We believe it is imperitive for a business owner to outline his / her vision and business needs before applying for a particular business loan and we wish you all the very best in your efforts to take your business to new heights as it so rightly deserves.

Chapter 16

Equipment

A major decision when developing a food product and its subsequent production is the kind of equipment needed to make it. It is important to have some knowledge about the various equipment available and how best they can be used in the process.

The pieces of equipment are either:

Hand tools: These can literally be held and operated by hand, and usually only one hand at that.

Utensils: These include implements, specialised tools, vessels (pots, pans and mixing bowls) that one uses in a kitchen. Many of these require more than one hand to operate

Broadly speaking equipment can be divided on the basis of:

1. **Size**: First there is small equipment, things that are easy to carry and move about the kitchen. Then there is large equipment, these items are usually immobile and are often fixed to the floor or a bench.

2. **Order of use:** This is dictated by the sequence they are used in while processing and preparation. They can be classified as:

   a. *Receiving equipment*: Receiving platforms, weighing scales, etc

   b. *Storage equipment*: Jars, bottles, shelves, bins, racks, etc used to store food.

   c. *Kitchen equipment*: These include all tools and utensils used for preparation, cooking, holding, serving and washing procedures

3. **Mode of use**: Based on this criteria they can be classified as non-mechanical, mechanical and powered.

   a. *Non-mechanical*: Non-mechanical equipment generally does not have any moving parts and is usually small. Most hand-held tools and utensils fall into this category that includes knives, mixing bowls and spoons.

   b. *Mechanical:* Because many kitchen tasks are time consuming and repetitive, many kitchens also have a variety of mechanical devices in order to make these tasks easier. Most of these pieces of equipment are small enough to either be held in the hand or moved around freely.

   c. *Powered:* Finally there is powered equipment. These pieces of equipment need electricity, gas or steam to work. Some power equipment, like electric knives, are small and are hand-held. Others, like ovens, are so large and heavy that they are fixed in one spot. Most powered equipment with moving parts, like mixers and blenders, are driven by electricity. Ovens, boilers and steamers, which have a few moving parts but generate heat, are usually powered by gas, wood or electricity.

Improvements in the design and manufacture of powered equipment have meant that one person can do the work of several people in a very short time. Also some tasks are performed better by machine than could ever be done by hand.

Equipment Safety

Many pieces of kitchen equipment use a chopping or mincing action with sharp blades moving at high speed and can cause serious injury if not used properly. One needs to be extra careful and follow the proper health and safety regulations when using this type of equipment.

Before using any piece of kitchen equipment, one should spend some time reading through the operator's manual. The manual will not only contain information about how to use the equipment but will also tell how to clean and maintain the equipment.

Equipment Chart

The following chart is a useful way to categorise the different types of kitchen equipment.

Non-mechanical			
Small items		Large items	
Hand tools	Utensils	Mobile and bench-top	Fixed or immobile
Knives	Mixing bowls	Stock pot	Benches
Whisk	Chinoise		Shelving
Piping bag	Colander		
Piping nozzles	Sieve		
Scraper	Ladle		
Wooden spoons	Spoons		
Plain rolling pin	Saucepans		
Measuring spoons	Fry pans		
	Tongs		
	Scoops		
	Fish slice		
	Spiders		
	Measuring jugs		
	Grater		
	Oven trays and dishes		

Mechanical			
Small items		Large items	
Hand tools	Utensils	Mobile and bench-top	Fixed or immobile
Olive stoner	Rolling pin	Pan scales	Floor scales
Egg slicer	Can opener		
Hand juicer	Scales		
Kitchen scissors	Vegetable mill		
Potato scoop	Timer		
	Mandolin		
	Thermometer		
	Mortar and pestle		

Small items		Large items	
	Powered:		
	gas, steam or electric		
Hand tools	*Utensils*	*Mobile and bench-top*	*Fixed or immobile*
Wand-type blenders	Scales	Electric mixer	Refrigerators
	Blender	Potato peeler	Freezers
Electric knives	Food processor and attachments	Bowl chopper	Boiler
		Electric slicing machines	Brat pan
	Electric mixer		Dishwasher
	Timer	Deep fryer	Steamer
	Thermometer	Microwave	Stove
		Toaster	Ovens
			Electric mixer
			Fridge
			Bain-Marie
			Char grill
			Flat top grill
			Salamander
			Hot press
			Deep fryer

Selection Criteria for Equipment

The various pieces of equipment are selected for a specific purpose based on the following factors:

1. Size of business plant
2. Type of business plant
3. Food product produced
4. Usage
5. Utility
6. Functionality
7. Frequency of use
8. Price
9. Ease of installation
10. Ease of maintenance
11. Ease of operation
12. Safety
13. Economy

14. Ease of cleaning
15. Attractiveness
16. Source of supply
17. Quality control
18. Warranty and guarantee terms
19. Local service centre
20. Efficiency

Design, Construction and Installation

All equipment and utensils are designed, constructed and installed to function as intended, to permit effective cleaning and sanitation and to prevent contamination of food products

o Equipment is designed, constructed and installed to ensure that:
  • the process is capable of delivering the results which are anticipated (e.g. effective package sealing, control of gas flush composition and time, where MAP is used);
  • it can be adequately and easily cleaned, sanitized, maintained and inspected to prevent contamination of the product during operations;
  • contamination of the product during operation is prevented (e.g. location of lubricant reservoirs);
  • equipment is exhausted to the outside to prevent excessive condensation where necessary; and
  • proper drainage is permitted and where appropriate, equipment is connected directly to drains. Where applicable, drains are fitted with backflow preventers.

Food Contact Surfaces

Food contact surfaces are constructed of appropriate materials and are maintained in a manner to prevent contamination of food.

o Food contact surfaces of equipment, containers and utensils are smooth, non-corrosive, non-absorbent, non-toxic, free from pitting, cracks or crevices, and able to withstand repeated cleaning and sanitation.

o When coatings, paints, chemicals, lubricants and other materials are used for food contact surfaces or utilized on equipment where there is a possibility of contact with food, the substances are appropriate for the intended use and are used in accordance with the manufacturer's instructions.

o Equipment and utensils used to handle inedible materials are not used to handle edible material.

Equipment Maintenance and Calibration Program

An effective maintenance and calibration program is in place to ensure that equipment performs consistently as intended and prevents contamination of the product.

- The manufacturer has an effective written preventive maintenance and calibration program to ensure that equipment which may impact on food safety functions as intended. This includes:
  - o a list of equipment requiring regular maintenance; and
  - o maintenance procedures and frequencies (e.g. equipment inspection instructions, a schedule of adjustments and part replacements based on the equipment manufacturer's manual or equivalent or based on operating conditions that could affect the condition of the equipment).
- The manufacturer establishes written protocols, including calibration methods and frequencies, for equipment monitoring and/or controlling devices that may impact on food safety.
- Equipment is maintained in a manner which ensures that there is no potential for the development of physical or chemical hazards (e.g. hazards resulting from inappropriate repairs, flaking paint and rust, excessive lubrication).
- Maintenance and calibration of equipment are performed by appropriately trained personnel.
- When routine or emergency repairs are made to equipment, in direct or indirect contact with food, an inspection to assess the compliance of the repair is performed before the equipment is used.
- The preventive maintenance and calibration programs and associated written protocol are followed.

Instrumentation Maintenance and Calibration Program

Instrumentation is designed, constructed, installed, calibrated and maintained such that the equipment is capable of delivering the required process, thereby ensuring product safety.

Improper design, installation, calibration or maintenance of instruments can lead to inadequate processing of the product, misuse of food additives, nutritional inaccuracies or composition violations.

- o The manufacturer has an effective written preventive maintenance and calibration program to ensure that instrumentation which may impact on food safety functions as intended. This includes:
  - a list of instrumentation requiring regular maintenance and calibration; and
  - the maintenance and calibration procedures and frequencies
- o Instruments which control factors that may have an impact on food safety are designed, installed, constructed, calibrated and maintained as necessary to ensure that they function as intended.
- o Maintenance and calibration of instrumentation are performed by appropriately trained personnel.
- o Preventive maintenance, calibration programs and associated written protocols are followed.

The following are some examples of instrumentation that may be required to control factors significant to the process:

Temperature Measuring Devices

- o The manufacturer uses one temperature scale consistently throughout the processing system (e.g. Celsius or Fahrenheit).
- o Temperature measuring devices are calibrated against a known standard just prior to installation, and a minimum of once per year thereafter (or more frequently as recommended in the equipment manufacturer's manual), and are maintained as necessary to ensure accuracy.

Temperature Recorders

- o The scale of the temperature recording chart is not more than 12°C/cm (55°F/in) within the range of 10°C (18°F) of process

temperature, and the chart graduation does not exceed 1°C (2°F) within 6°C (11°F) of processing temperature.
o The accuracy of temperature recorders is verified upon installation, and thereafter, a minimum of once per year (or more frequently as necessary to ensure their accuracy).

Timing Devices
o Timing devices and recorders are verified upon installation, and thereafter annually (or more frequently as necessary to ensure accuracy).
o Where timing devices are not equipped with a power backup, controls are in place to verify that process time requirements are met.
o Any official timing device is located so that it can be easily and accurately read by the operators.

Pressure Gauges
o Each pressure gauge is calibrated at least annually or more frequently as necessary to ensure accuracy.

Metal Detectors
o Metal detection equipment is designed, constructed, installed, calibrated and maintained in accordance with the equipment manufacturer's manual, to ensure effective removal of metals. This may include adjustment for product effect, selection of target metal and size, timing of the reject mechanism and suitability for environmental conditions.

Scales/Metering Devices
o The sensitivity is appropriate to the use.
o Scales are designed and installed to withstand the environmental conditions or are adequately protected (e.g. away from drafts, rust, corrosion, etc.).
o Scales and meters are calibrated in accordance with the equipment manufacturer's manual to ensure accuracy at all times.

Other Instrumentation

o Other specialized instrumentation when used to control factors significant to food safety, are calibrated as necessary.

Space

Building Exterior

Outside Property and Buildings

Buildings and surrounding areas are designed, constructed and maintained in a manner which prevents conditions which may result in the contamination of food.

Grounds, Roadways and Drainage

o The surrounding land is maintained to control sources of contamination such as debris and pest harbourage areas.
o The building is not located in close proximity to any environmental contaminants.
o Roadways are properly graded, compacted, dust proofed and drained.
o The surrounding property is adequately drained.

Exterior Building Structure

o The building exterior is designed, constructed and maintained to prevent entry of contaminants and pests. For example, the exterior has no unprotected openings; air intakes are appropriately located; and the roof, walls and foundation are maintained to prevent leakage.

Building Interior

Design, Construction and Maintenance

Building interiors and structures are designed, constructed and maintained to prevent conditions which may result in the contamination of food.

Floors, Walls and Ceilings

o Floors, walls and ceilings are constructed of materials that are durable, impervious, smooth, cleanable, and suitable for the production conditions in the area (e.g. materials will not result in the contamination of the environment or food).

o Where appropriate, wall, floor and ceiling joints are sealed and angles are coved to prevent contamination and facilitate cleaning.

o Floors are sufficiently sloped to permit liquids to drain to trapped outlets.

o Ceilings, overhead structures, stairs and elevators are designed and constructed to prevent contamination.

o Floors, walls, ceilings and all overhead structures are maintained to prevent deterioration (e.g. rust, flaking paint) and contamination (e.g. dust, mould).

Windows and Doors

o Windows are sealed or equipped with close-fitting screens.

o Where there is a likelihood of breakage of glass windows that could result in the contamination of food, the windows are constructed of alternative materials or are adequately protected.

o Doors have smooth, non-absorbent surfaces and are close-fitting and self-closing where appropriate to prevent the entry of pests and vermin.

Process Flow Separation

o Buildings and facilities are designed to facilitate hygienic operations (e.g. there is regulated flow in the process, from the arrival of the raw material at the premises to the finished product).

o Activities are adequately separated by physical or other effective means where cross-contamination (e.g. biological, chemical [e.g. allergens], physical) may result. For example:

• all untreated raw ingredients or materials are kept separate from in line or finished products to prevent microbiological cross-contamination.

• potential allergens are controlled to avoid the possibility of cross-contamination.

o It is preferred that dedicated processing lines and utensils are used for allergen containing products.

o Where common pieces of equipment are used for the processing of allergen containing products and non-allergen containing products, the manufacturer has established a procedure that does not result in the contamination of food products with allergens.

o The manufacturer has established a process flow from incoming materials to finished products with no cross-over or areas of concern that can cause undeclared allergens to be present in the finished products (e.g. conveyor belts, shared equipment).

o Where mobile equipment (such as fork lifts or equivalent) moves between incompatible areas, measures are taken to minimize cross-contamination.

o Pallet's design, condition and use are specified to avoid contamination.

Lighting: Lighting is adequate for the activity being conducted. Where appropriate, light bulbs and fixtures are protected to prevent contamination of food or packaging material.

o Lighting is appropriate such that the intended production or inspection activity can be effectively conducted. The lighting does not alter food colour and is not be less than the following:

- 540 lux (50 foot candles) in inspection areas (inspection areas are defined as any point where the food product or container is visually inspected or instruments are monitored)

- 220 lux (20 foot candles) in work areas

- 110 lux (10 foot candles) in other areas

o Light bulbs and fixtures located in areas where there is exposed food or packaging material are of a safety type or are protected to prevent the contamination of food or packaging material in case of breakage (e.g. shatterproof bulbs or bulb covers).

Ventilation: Adequate ventilation is provided to prevent excessive heat, steam, condensation and dust, as well as to remove and minimize entry of contaminated air. Air used for processing techniques is appropriately sourced and treated.

○ Ventilation provides sufficient air exchange to prevent unacceptable accumulations of steam, condensation, dust or excessive heat and to minimize entry of contaminated air.

○ Ventilation systems are constructed to avoid air flow from less clean areas (e.g. the receiving area) to clean areas (e.g. packaging and finished product storage) and designed to be adequately maintained and cleaned.

○ Ventilation openings are equipped with close-fitting screens or filters as appropriate to prevent the intake of contaminated air (dust, dirt) or entry of insects and rodents. Air filters (e.g. filters for intake air and compressed air) are checked, cleaned or replaced at least as often as the manufacturer specifies or more frequently if a problem is indicated (e.g. evidence of filter fouling or perforation).

○ Air used as a processing technique (e.g. air blows, air dryers, etc.) is appropriately sourced and treated (air intakes, filters, compressors) to reduce any source of contamination.

Waste disposal: Sewage, effluent and waste storage and disposal systems are designed, constructed and maintained to prevent contamination.

○ Drainage and sewage systems are equipped with appropriate traps and vents.

○ Establishments are designed and constructed so that there is no cross-connection between the sewage system and any other waste effluent system in the establishment.

○ Effluent or sewage lines do not pass directly over or through production areas unless they are controlled to prevent contamination.

○ Adequate facilities and equipment are provided and maintained for the storage of waste and inedible material prior to their removal from the establishment. These facilities are designed to prevent contamination.

○ Containers used for waste are clearly identified, leakproof, moisture resistant, easy to clean and, where appropriate, covered.

o Waste is removed and containers are cleaned and sanitized at an appropriate frequency to minimize the potential for contamination.

Sanitary Facilities

Employee Facilities: Employee facilities are designed, constructed and maintained to permit effective employee hygiene and to prevent contamination.

o Processing areas are provided with an adequate number of conveniently located handwashing stations (preferably hands-free) with trapped waste pipes to drains.

o Washrooms, lunchrooms and change rooms are adequately ventilated and maintained in clean condition. They are separate from and do not lead directly into food processing areas.

o Washrooms have handwashing facilities with a sufficient number of maintained sinks that are properly trapped to drains.

o Handwashing facilities are adequately maintained and have hot and cold running potable water distributed from a single nozzle, soap, sanitary hand-drying supplies or devices, and, where required, a cleanable waste receptacle.

o Handwashing stations, hand dips and footbaths are maintained in all applicable areas of the facility.

o Notices to wash hands are posted in appropriate areas.

Equipment cleaning and sanitizing facilities: Facilities for cleaning and sanitizing equipment are adequately designed, constructed and maintained to prevent contamination.

o Facilities are constructed of corrosion resistant materials which are capable of being easily cleaned, and are provided with potable water at temperatures appropriate for the cleaning chemicals used.

o Equipment cleaning and sanitizing facilities are adequately separated from food storage, processing and packaging areas, to prevent contamination.

Water/Ice/Steam Quality

Water and Ice: The potability of hot and cold water is controlled to prevent contamination.

o Potable water meets the requirements of Health Canada's *Guidelines for Canadian Drinking Water Quality* and any applicable provincial and municipal requirements.

o Water is analysed by the fresh alimentary paste manufacturer at a frequency adequate to confirm its potability. For microbial analysis, water from a municipal water source is analysed semi-annually and water from other sources is analysed on a monthly basis. For chemical analysis, water from non-municipal sources is adequately analysed at least at the initial start-up of the well.

o The fresh alimentary paste manufacturer has contingency plans in place to deal with provincial/municipal orders to boil water and unsatisfactory water analysis results.

o There are no cross-connections between potable and non-potable water supplies. All hoses, taps and other similar sources of potential contamination are designed to prevent back-flow or back siphonage.

o Water treatment chemicals, where used, are appropriate for the intended use and are used in accordance with the chemical manufacturer's instructions.

o The chemical treatment is monitored and controlled to deliver the desired concentration and to prevent contamination.

o Recirculated water is treated, monitored and maintained as appropriate to the intended purpose. Recirculated water has a separate distribution system, which is clearly identified.

o Where filters are used they are kept effective and maintained in a sanitary manner.

o In areas for food processing, handling, packaging and storage, water temperatures and pressures are adequate for all operational and clean-up needs.

o Ice used as an ingredient or ice used in direct contact with food is made from potable water and is protected from contamination.

Steam: The potability of steam which is in direct contact with food or food contact surfaces is controlled to prevent product contamination. Steam supply is adequate to meet operational requirements.

o Boiler treatment chemicals used are appropriate for the intended use and are used in accordance with the chemical manufacturer's instructions.

o Boiler feed water is tested regularly and the chemical treatment is controlled to prevent contamination.

o The steam supply is generated from potable water and is adequate to meet operational requirements.

o Traps are provided as necessary to ensure adequate condensate removal and elimination of foreign materials.

Chapter 17

Setting a Price for the Product

Definition: *To establish a selling price for a product.*

"It's probably the toughest thing there is to do," says Charles Toftoy, Associate Professor of management science at George Washington University. "It's part art and part science."

No matter what type of product sold, the price charged to customers or clients will have a direct effect on the success of the business.

PRODUCT COST

Product cost refers to the costs used to create a product. These costs include direct labor, direct materials, consumable production supplies, and factory overhead.

Product cost can also be considered the cost of the labor required to deliver a service to a customer. In the latter case, product cost should include all costs related to a service, such as compensation, payroll taxes, and employee benefits.

The cost of a product on a unit basis is typically derived by compiling the costs associated with a batch of units that were produced as a group, and dividing by the number of units manufactured. The calculation is:

$$\text{Product unit cost} = \frac{\text{Total direct labor + Total direct materials +}}{\text{Total number of units}}$$

Though pricing strategies can be complex, the basic rules of pricing are straightforward:

○ All prices must cover costs and profits.
○ The most effective way to lower prices is to lower costs.
○ Review prices frequently to assure that they reflect the dynamics of cost, market demand, response to the competition, and profit objectives.
○ Prices must be established to assure sales.

Methods of Establishing Prices

Prices are generally established in one of four ways:

1. **Cost-plus pricing**: Many manufacturers use cost-plus pricing. The key to being successful with this method is making sure that the "plus" figure not only covers all overhead but generates the percentage of profit required. If the overhead figure is not accurate, it risks profits that are too low.

   A simple formula for this is as follows:

   Required sale Price = Cost of materials + Cost of labour + Overheads + Desired profit on sales.

2. **Demand price:** Demand pricing is determined by the optimum combination of volume and profit. Products usually sold through different sources at different prices—retailers, discount chains, wholesalers, or direct mail marketers—are examples of goods whose price is determined by demand.

   A wholesaler might buy greater quantities than a retailer, which results in purchasing at a lower unit price. The wholesaler profits from a greater volume of sales of a product priced lower than that of the retailer. The retailer typically pays more per unit because he or she are unable to purchase, stock, and sell as great a quantity of product as a wholesaler does. This is why retailers charge higher prices to customers.

3. **Competitive pricing:** Competitive pricing is generally used when there is an established market price for a particular product or service. If all the competitors are charging "X" for a Yoggurt drink, for example, that is what this product should also cost.

Competitive pricing is used most often within markets with commodity products, those that are difficult to differentiate from another.

If there is a major market player, commonly referred to as the market leader, that company will often set the price that other, smaller companies within that same market will be compelled to follow.

o To use competitive pricing effectively, know the prices each competitor has established. Then figure out the optimum price and decide, based on direct comparison, whether it can be defended.

o Should you wish to charge more than the competitors, there should be a case for a higher price, such as providing a superior customer service or better ingredients.

4. **Markup pricing:** Used by manufacturers, wholesalers, and retailers, a markup is calculated by adding a set amount to the cost of a product, which results in the price charged to the customer. For example, if the cost of the product is "X" and the selling price is "Y", the markup would be " Y–X".

This pricing method often generates confusion—not to mention lost profits—among many first-time small-business owners because markup (expressed as a percentage of cost) is often confused with gross margin (expressed as a percentage of selling price).

TERMS AND DEFINITIONS

Overhead Expenses

Overhead refers to all non-labor expenses required to operate the business. These expenses are either fixed or variable:

o **Fixed expenses.** No matter what the volume of sales is, these costs must be met every month. Fixed expenses include rent or mortgage payments, depreciation on fixed assets (such as cars and office equipment), salaries and associated payroll costs, liability and other insurance, utilities, membership dues and subscriptions (which can sometimes be affected by sales volume), and legal and accounting costs. These expenses do not change, regardless of whether a company's revenue goes up or down.

- o **Variable expenses.** Most so-called variable expenses are really semi-variable expenses that fluctuate from month to month in relation to sales and other factors, such as promotional efforts, change of season, and variations in the prices of supplies and services. Fitting into this category are expenses for telephone, office supplies (the more business, the greater the use of these items), printing, packaging, mailing, advertising, and promotion. When estimating variable expenses, use an average figure based on an estimate of the yearly total.

Cost of goods sold: Cost of goods sold, also known as cost of sales, refers to the cost to purchase products for resale or to the cost to manufacture products.

Freight and delivery charges are customarily included in this figure.

Accountants segregate cost of goods on an operating statement because it provides a measure of gross-profit margin when compared with sales, an important yardstick for measuring the business' profitability. Expressed as a percentage of total sales, cost of goods varies from one type of business to another.

Determining margin: Margin, or gross margin, is the difference between total sales and the cost of those sales. For example: If total sales equals "X" and cost of sales equals "Y", then the margin equals "X–Y".

At the same time, be aware of the risks that accompany making poor pricing decisions. There are two main pitfalls that can be encountered—underpricing and overpricing.

- o **Underpricing.** Pricing the products for too low a cost can have a disastrous impact on the bottom line, even though business owners often believe this is what they ought to do in a down economy.

    "Accurately pricing your product is critical at any point in the economic cycle but no more so than in a recession," says Laura Willett, a small business consultant and faculty member in the finance department at Bentley College in Waltham, Mass. "Many businesses mistakenly under price their products attempting to convince the consumer that their product is the least expensive alternative hoping to drive up volume; but more often than not it is simply perceived as cheap."

"Reducing prices to the point where you are giving away the product will not be in the firm's best interest long term," Willett says.

○ **Overpricing.** On the flip side, overpricing a product can be just as detrimental since the buyer is always going to be looking at your competitors pricing, Willett says. Pricing beyond the customer's desire to pay can also decrease sales. Toftoy says one pitfall is that business people will be tempted to price too high right out of the gate. "They think that they have to cover all the expenses of people who work for them, the lease, etc. and this is what price it takes to do all that," he says. "Put yourself in the customer's shoes. What would be a fair price to you?" He advises taking a little surveys of customers with two or three questions on an index-card-sized form, asking them whether the pricing was fair.

Factors to consider while pricing the product

a. Know the customer
b. Know the costs
c. Know the revenue target
d. Know the competition
e. Know where the market is headed
f. Have a budget action plan in place.

The product price should vary depending on a number of factors including:

○ **What the market is willing to pay.**
○ **How the company and product are perceived in the market.**
○ **What the competitors charge.**
○ **Whether the product is "highly visible" and frequently shopped and compared.**
○ **The estimated volume of product one can sell.**

A STEP BY STEP APPROACH TO COSTING OR PRICING OF A PRODUCT

Calculate the cost of running the business. A basic pricing method requires that one determines the full cost of running the business and price the product in such a way as to keep the business in the black. So, the first thing one needs to do is calculate how much it costs to run the business. These costs can be further divided into direct and indirect costs.

o **Direct costs** are those which are immediately associated with doing business. These costs get directly assigned to the products and services provided.
o Labour costs
o Marketing costs
o Manufacturing costs (cost of raw materials, equipment, etc.)

o **Indirect costs** are things associated with keeping things running, on the day to day. These are sometimes thought of as the hidden or even "true cost" of running a business.
o Operating expenses (including rent on the building, utilities, etc.)
o Debt service costs
o Return on any investment capital
o Cleaning and office supplies
o Salary

Set a "success point." The only reason to start a business is to make money, and specifically to make enough money to keep the venture a successful enterprise.

Anticipate the customer's desires. Identify the customer base and their buying tendencies.
o How much do they desire the particular product?
o Is there a demand for it?
o Be as specific as possible in the discussion of numbers.
o How much is it possible to sell, given the current resources?
o How much does one needs to sell to maintain the visibility and success of the current model? What might need to be changed?

Study the Competition

o **Understand the effects of over and underpricing.** Setting the price inefficiently will have marked and measurable effects on the numbers. One needs to learn to recognize the symptoms of having either a low or high price point. This can indicate that one may need to make a change.

o **Keep a close eye on the pricing and the budget.** Monitor the profits and prices at least monthly. Break down the cost/gains of every product so that one knows how each contributes to the overall profitability month-to-month. This can give a clear picture of the money flow.

o **Develop a budget plan.** Try to focus on a longterm strategy that will result in making the business profitable. This might not involve making drastic changes right away, but slowly moving toward an overall goal of profitability.

o **Raise prices slowly and incrementally.** Sudden increase will look like desperate moves made by a struggling business, which may or may not be true. Watch the sales volume immediately after making the change. If the move was too sudden, one will see a negative change, suggesting that one needs to do more to sell the new variation on the product and justify its price.

o **Use promotions to lower prices and get people in the store.** Using promotions for limited periods of time, or coupons that expire, you can help drive customers toward a particular product or service.

o **Use creative promotions to get people in the door.** Use a Buy One, Get One Free promotion.

o **Appeal to the customer's emotions and rationality.** Promotional pricing strategies cannot just be informational campaigns, they have to connect with the target market. In order to do this, take time to appeal to their emotions or pragmatism. A common business strategy involves pricing items in .99 cent increments, rather than dollar increments.

o **Try to up-sell promotions to move more merchandise.** In Optional Product pricing, companies will attempt to increase the amount customers spend once they start to buy. Optional 'extras' increase the overall price of the product or service.

o **Avoid the appearance of price gouging.** Gouging involves raising the product to a high price because one has a substantial competitive advantage of some kind, or a corner of the market. This advantage is not sustainable. The high price tends to attract new competitors into the market, and the price inevitably falls due to increased supply.

- *Captive product pricing* is used when products have complements. Companies will charge a premium price where the consumer is captured.

IN A NUTSHELL

Step 1: Evaluate the pricing environment
Step 2: Define the pricing objectives
Step 3: Choose a pricing strategy
Step 4: Formulate pricing tactics
Step 5: Assemble a pricing mix

Chapter 18

Advertising and Marketing

The success of any food product is measured by its sales and popularity amongst the consumers. A pivotal aspect of how to make a product successful lies in the two pillars, viz. advertising and marketing. These are capable of swaying public opinion to a large extent.

Though often used interchangeably, the two are quite different in their approach.

To put it simply, advertising is just one component, or subset, of marketing. Public relations, media planning, product pricing and distribution, sales strategy, customer support, market research and community involvement are all parts of comprehensive marketing efforts.

Advertising: The paid, public, non-personal announcement of a persuasive message by an identified sponsor; the non-personal presentation or promotion by a firm of its products to its existing and potential customers.

Marketing: The systematic planning, implementation and control of a mix of business activities intended to bring together buyers and sellers for the mutually advantageous exchange or transfer of products.

The best way to distinguish between advertising and marketing is to think of marketing as a pie, inside that pie you have slices of advertising, market research, media planning, public relations, product pricing, distribution, customer support, sales strategy, and community involvement.

Many marketing departments are, whether by choice or design, insulated from other business functions such as sales and customer support. It is this separation that causes a disconnect between a company's intended aim of getting new customers, and the actual follow through.

All marketing elements must work independently, as well as interdependently.

According to Kathleen Micken, Assistant Professor of marketing for the Gabelli School of Business at Roger Williams University, "Marketing might be defined as everything an organization does to facilitate an exchange between itself and its customers/clients. Advertising is just one of many marketing activities." Steven R Jolly, owner of SRJ Marketing Communications, a marketing and design firm in Dallas, Texas adds, "Marketing is the sum total of all impressions and advertising is part of the impressions that must be managed. And, of course, advertising has a hard dollar cost associated".

ADVERTISING

Advertising has become an increasingly important tool for food manufacturers to sell their products. No longer it is sufficient for a food product to be wholesome, good tasting, and available on store shelves. With the proliferation of new food products responding to increasingly diverse consumer demands and an expanding diversity in channels of distribution, advertising has become critical to the success of new products and the survival of old favourites. Consumers need to know about the existence of a product, understand how it meets their needs, and know where to obtain it. Launching a new product without a well-designed and executed advertising program may condemn an otherwise excellent product to an early trip to the remainder aisle. Importantly, not only must an advertising program be effective, it must be legally sound in order to avoid a potentially serious detour to the courthouse or regulatory agency.

Advertising is a single component of the marketing process. It is the part that involves getting the word out concerning your business, product, or the services you are offering. It involves the process of developing strategies such as ad placement, frequency, etc. Advertising includes the placement of an ad in such mediums as newspapers, direct mail, billboards, television, radio, and of course the Internet.

Advertising is the largest expense of most marketing plans, with public relations following in a close second and market research not falling far behind.

Advertising, according to Barron's Dictionary of Marketing Terms, is the "paid form of a nonpersonal message communicated through various media. [It] is persuasive and informational and is designed to influence the purchasing behavior and/or thought patterns of the audience".

According to Donna P Anderson, APR, marketing and practice development director for Andrews & Kurth LLP, an international law firm, "Advertising is a tactic, or specific activity conducted to implement a strategic marketing or public relations plan. Advertising may be one of the tools used to meet a goal."

Advertising includes direct mail, newspapers, magazines, television, radio, Internet and out of house (billboards). Advertising is any direct or indirect communication of information by a company about itself or its products. Magazine ads, Web pages, and television commercials are obvious forms of advertising. Less obvious forms of advertising include the sponsorship of a cooking school, paying actors to use certain products in television shows or movies, and press information kits distributed to food editors of local newspapers. Advertising also includes written materials designed for use by distributors, retailers, or brokers (so-called sell sheets); statements made by servers offering samples at the local club store; and, of course, claims made on product labels.

When selecting the best advertising venues, one will need to consider budget, target audience and message.

A "media mix" is almost always necessary to get the penetration one needs—a mix of radio, television and direct mail, for example. Because the prospective client is bombarded by thousands of ads each day, one cannot depend on just one advertising vehicle.

Frequency is also an important part of any advertising campaign—if running print ads in a monthly publication, one will need to augment the monthly buy with more regular ads, say in a weekly or daily publication, television or radio.

To select the right media mix, carefully narrow down the target audience. Use media that targets the primary prospect group, and then hit the same demographic over and over again to stay top of mind.

When preparing the cost analysis of various media outlets, include a cost per thousand (CPM) column. Cost per thousand tells how much it costs to reach 1000 people.

Food advertising is all about presenting the most delectable preparations though in reality it might be a very simple meal, with the intention of tempting people to buy.

A successful advertisement creates a desire in viewers, listeners or readers. It also provides information on how to fulfill that desire and makes the potential customer feel good about doing so. With so many products and service providers in the marketplace, using a proven technique in the advertising increases the likelihood that it will return value.

Basic Techniques used in Advertising

Repetition

Repetition is a simple yet effective technique used to build identity awareness and customer memory. Even advertisements using other successful approaches mention the product or company name more than once, particularly in television because its combination of sight and sound, allows the advertiser to disguise the repetition by changing its delivery (from visual to audio).

Claims

Advertising that promotes specific features or makes claims about what a product or service can do for the potential customers provides successful results by informing, educating and developing expectations in the buyer. Claims can state facts or simply use hype, such as calling one brand of orange juice "the best" when nutritionally it is identical to other brands. Claims may mislead through omission or by using what some advertisers and political campaigners call "weasel words." These are subtle statement modifiers that render the claim meaningless if studied closely. Common weasel words include "helps," "fights" and "virtually."

Association

Associating a product or company with a famous person, catchy jingle, desirable state of being or powerful emotion creates a strong psychological connection in the customer. These ads encourage an emotional response in customers, which then is linked to the product being advertised, making it attractive through transference.

Bandwagon

The bandwagon technique sells a product or service by convincing the customer that others are using it and they should join the crowd. Other bandwagon advertisements suggest that the customer will be left out if they do not buy what is being sold.

Promotions

Coupons, sweepstakes, games with prizes and gifts with purchases create excitement, and participation encourages customers to build a relationship with the sponsoring product or service. The attraction of getting something "free" or earning "rewards" makes promotions successful.

Persuasive Techniques in Advertising

The persuasive strategies used by advertisers who want you to buy their product can be divided into three categories: Pathos, logos, and ethos.

Pathos: An appeal to emotion. An advertisement using pathos will attempt to evoke an emotional response in the consumer. Sometimes, it is a positive emotion such as happiness: An image of people enjoying themselves while drinking Pepsi. Other times, advertisers will use negative emotions such as pain: A person having back problems after buying the "wrong" mattress. Pathos can also include emotions such as fear and guilt: Images of a starving child persuade you to send money.

Logos: An appeal to logic or reason. An advertisement using logos will give the evidence and statistics needed to fully understand what the product does. The logos of an advertisement will be the "straight facts" about the product: One glass of *Real* orange juice contains 75% of daily vitamin C needs.

Ethos: An appeal to credibility or character. An advertisement using ethos will try to convince that the company is more reliable, honest, and credible; therefore, one should buy its product. Ethos often involves statistics from reliable experts, such as nine out of ten dentists agree that "A" is the better than any other brand..

Some common and most used techniques used by the advertisers to get desired results.

1. **Emotional appeal:** This technique of advertising is done with the help of two factors—needs of consumers and fear factor. Most common appeals under need are:
   a. Need for something new
   b. Need for getting acceptance
   c. Need for not being ignored
   d. Need for change of old things
   e. Need for security
   f. Need to become attractive, etc.

   Most common appeals under fear are:
   a. Fear of accident
   b. Fear of death
   c. Fear of being avoided
   d. Fear of getting sick
   e. Fear of getting old, etc.

2. **Promotional advertising:** This technique involves giving away samples of the product for free to the consumers. The items are offered in the trade fairs, promotional events, and ad campaigns in order to gain the attention of the customers.

3. **Bandwagon advertising:** This type of technique involves convincing the customers to join the group of people who have bought this product and be on the winning side.

For example, recent Pantene shampoo ad which says "15 crores women trusted Pantene, and you?"

4. **Facts and statistics:** Here, advertisers use numbers, proofs, and real examples to show how good their product works. Sometimes citing figures from National and International agencies.

5. **Unfinished ads:** The advertisers here just play with words by saying that their product works better but do not answer how much more than the competitor. For example, Lays—no one can eat just one or Horlicks more nutrition daily. The ads do not say who can eat more or how much more nutrition.

6. **Weasel words:** In this technique, the advertisers do not say that they are the best from the rest, but do not also deny. "Weasel words" are used to suggest a positive meaning without actually really making any guarantee. A scientist says that a diet product might help you to lose weight the way it helped him to lose weight. A dish soap leaves dishes virtually spotless. For example, Sunsilk Hairfall Solution—reduces hairfall. The ad does not say stops hairfall.

7. **Endorsements:** The advertisers use celebrities to advertise their products. The celebrities or star endorse the product by telling their own experiences with the product.

8. **Complementing the customers:** Here, the advertisers used punch lines which complement the consumers who buy their products. For example, Revlon says "Because you are worth it."

9. **Ideal family and ideal kids:** The advertisers using this technique show that the families or kids using their product are a happy go lucky family. The ad always has a neat and well-furnished home, well-mannered kids and the family is a simple and sweet kind of family.

10. **Patriotic advertisements:** These ads show how one can support their country while he uses their product or service.

11. **Questioning the customers:** The advertisers using this technique ask questions to the consumers to get response for their products.

12. **Transfer:** Positive words, images, and ideas are used to suggest that the product being sold is also positive. A textile manufacturer wanting people to wear their product to stay cool during the summer shows people wearing fashions made from their cloth at a sunny seaside setting where there is a cool breeze.

13. **Plain folks:** The suggestion that the product is a practical product of good value for ordinary people. A cereal manufacturer shows an ordinary family sitting down to breakfast and enjoying their product.

14. **Snob appeal:** The suggestion that the use of the product makes the customer part of an elite group with a luxurious and glamorous lifestyle. A coffee manufacturer shows people dressed in formal gowns and tuxedos drinking their brand at an art gallery.

15. **Avante garde:** The suggestion that using this product puts the user ahead of the times. A toy manufacturer encourages kids to be the first on their block to have a new toy.

16. **Magic ingredients:** The suggestion that some almost miraculous discovery makes the product exceptionally effective. A pharmaceutical manufacturer describes a special coating that makes pain reliever less irritating to the stomach than a competitor.

17. **Bribe:** This technique is used to bribe the customers with something extra if they buy the product using lines like "buy one shirt and get one free", or "be the member for the club for two years and get 20% off on all services".

18. **Surrogate advertising:** This technique is generally used by the companies which cannot advertise their products directly. The advertisers use indirect advertisements to advertise their product so that the customers know about the actual product. The biggest example of this technique is liquor ads. These ads never show anyone drinking actual liquor and in place of that they are shown drinking some mineral water, soft drink or soda.

Tricks for Food Advertising

Tempting pictures: The first that attracts a reader's attention is the picture of a mouth watering product. Hence food manufacturers have to ensure that they use the best photography skills and "dress up" the product to look tempting. There are now food "stylists" who are like make-up artists for food, transforming even the most insipid product into a tempting delicacy.

Catchy jingles/songs/commercials: Food and music have a good connection. A catchy jingle turns all eyes towards a commercial, and the music forces viewers to see the advertisement till the end. Repeated viewing will actually tempt them to buy the food product.

Labels on packages: Though traditionally not considered part of advertising, food labels are mandatory and it is an advertising gimmick to present information in a manner that the least beneficial ingredients appear in miniscule amounts.

Depicting fun: All advertisements both as commercials on the screen or as newsprint depict fun and joyful activities along with the food product, which appeal to individuals caught in stressful environments. The ambience provides the right impact.

Characters that appeal: Most food companies try to rope in a character like a movie star, a sports personality or a cartoon character since the masses relate to him, and his endorsement of a brand adds to the appeal.

Tempting pricing: Irrespective of the actual price, the advertised price is always one that actually tempts potential customers. The price may be of a very small quantity but it attracts attention and lures people to the counter.

Offer a FREE sample: Give potential customers a product for free so they can try out.

Food Advertising Regulation

Although advertising is a constitutionally protected form of commercial speech, it is subject to extensive regulation. Food products, in light of health and safety concerns associated with them, are subject to more regulatory oversight than other types of consumer products.

Advertising that is false or deceptive is not constitutionally protected and may be illegal. Furthermore, even advertising that is literally truthful may be restricted by the government—

and even banned in some contexts—if it is misleading due to its context or if the government has determined that the public well-being requires that certain types of claims be limited. In the US, the Federal Trade Commission Act (the FTC Act) and associated regulations are the federal government's primary weapon against false, deceptive, or unfair advertising. Section 5 of the FTC Act prohibits "unfair or deceptive acts or practices in or affecting commerce." Section 12 of the FTC Act deals specifically with the false advertising of "food, drugs, devices or cosmetics," and prohibits the dissemination of any false advertisement for the purpose of inducing, or which is likely to induce, directly or indirectly, any purchase of such items. "False advertisement" is defined as any advertisement, other than labelling, that is misleading in a material respect. In determining whether an advertisement is misleading, the Federal Trade Commission (FTC) will take into account not only representations made or suggested by statement, word, design, device, sound, or any combination thereof, but also the extent to which the advertisement fails to reveal material facts about the product or its performance. The Food and Drug Administration (FDA) regulates claims made on labels of food products while the Department of Agriculture regulates claims made on labels of meat and poultry products. Section 201(m) of the Food, Drug, and Cosmetic Act (the FD&C Act) defines labelling as "all labels and other written, printed or graphic matter upon any article or any of its containers or wrappers or accompanying such article." Under this formulation, labelling includes not only actual product labels, but also point-of-sale materials (such as display-ready units, "shelf-talkers," and retail handouts) that accompany food products. And in a controversial reach for expanded jurisdiction, the FDA recently took the position that if a product label refers consumers to a Website, then FDA's regulatory authority over labelling extends to claims made on that Website (see FDA Warning Letter to Ocean Spray Cranberries, Inc., Jan. 19, 2001).

Under Section 403 of the FD&C Act, food labelling that is false or misleading in any particular causes the product to be misbranded in violation of Section 301 of the FD&C Act. Note that this is a much more rigorous standard than under the FTC Act, which prohibits only advertising claims that are misleading "in a material respect." While this distinction had some significance in the past (one court said that unlike the standard used by the FTC, FDA's regulatory authority was intended to protect "the ignorant, the unthinking and the credulous"), since December 2002 FDA has followed the FTC's "reasonable consumer standard" in determining whether a claim on a conventional food is misleading. FDA has adopted a comprehensive regulatory framework governing nutrition and health claims made on food products. FDA both restricts the number and type of nutrition and health claims that can be made on product labels and has issued detailed definitions and guidelines that must be met with respect to those claims that are allowed. FDA has adopted a highly technical structure characterizing different types of claims:

- **Health claims:** As defined by regulation, a health claim is any claim that expressly or by implication characterizes the relationship of any substance to a disease or health-related condition (for example, a claim that links the consumption of soluble fiber to a reduced risk of heart disease is considered by FDA to be a health claim). Health claims can only be made on food labels if supported by the "totality of publicly available scientific evidence" and there must be "significant scientific agreement" among qualified experts that the claim is supported by such evidence. FDA has authorized only a limited number of health claims. See 21 C.F.R. §§ 101.70-101.83.

- **Qualified health claims:** Qualified health claims are claims that describe substance/disease relationships based on "competent and reliable scientific evidence" but that must be accompanied by explicit qualifying language informing consumers that the evidence supporting the claim is not conclusive.

- **Structure/function claims**: Structure/function claims describe the effect of a particular food or nutrient on the structure or function of the body (for example, the claim that a particular food or nutrient helps maintain joint health and flexibility).

- **Dietary guidance statements:** A dietary guidance statement addresses the role that general categories of food play in the diet. For example, "Diets rich in fruits and vegetables may reduce the risk of some types of cancer and other chronic diseases."

MARKETING

Marketing is actually much more than simply advertising; it is everything done to promote the business and food products, from the moment it is conceived to the point at which customers buy it. It involves researching, selling, distributing, and promoting the product.

Marketing is defined as the action or business of promoting and selling products or services, including market research and advertising.

Food marketing brings together the food producer and the consumer through a chain of marketing activities.

Pomeranz & Adler, 2015, define food marketing as a chain of marketing activities that take place within the food system between a food organisation and the consumer.

This has the potential to be a complicated procedure, as there are many processes that are used prior to the sale of the food product. These include food processing, wholesaling, retailing, food service and transport. On a global scale, the food marketing industry is one of the largest direct and indirect employers.

For Schaffner & Schroder, 1998, food marketing is the act of communicating to the consumer through a range of marketing techniques in order to add value to a food product and persuade the consumer to purchase. This includes all activities that occur in between the completion of a product and the purchasing process of consumers.

Food marketing systems differ worldwide due to the level of development in the particular country, economically and technologically.

Marketing Mix

The four components of food marketing are often called the "four Ps" of the marketing mix because they relate to product, price, promotion, and place. **(Discussed in detail in Chapter 2.)**

The money that manufacturers invest in developing, pricing, promotion, and placing the products; differentiate a food product on the basis of both quality and brand-name recognition.

Product

In deciding what type of new food products a consumer would most prefer, a manufacturer can either try to develop a new food product or try to modify or extend an existing food. For example, a flavoured "srikhand" would be a new product, but milk in a new flavour would be an extension of an existing product.

There are three steps to both developing and extending: Generate ideas, screen ideas for feasibility, and test ideas for appeal. Only after these steps will a food product make it to national market.

The food industry faces numerous marketing decisions. Money can be invested in brand building (through advertising and other forms of promotion) to increase either quantity demanded or the price consumers are willing to pay for a product.

Coca Cola, for example, spends a great deal of money both on perfecting its formula and on promoting the brand. This allows Coke to charge more for its product than can makers of regional and smaller brands.

Manufacturers may be able to leverage their existing brand names by developing new product lines. For example, Heinz started out as a brand for pickles but branched out into ketchup.

The consumer, to be brand loyal, must be able to actively resist promotional efforts by competitors. A brand loyal consumer will continue to buy the preferred brand even if a competing product is improved, offers a price promotion or premium, or receives preferred display space. Some consumers have multi-brand loyalty. Here, a consumer switches between a few preferred brands. The consumer may either alternate for variety or may, as a rule of thumb, buy whichever one of the preferred brands is on sale. This consumer, however, would not switch to other brands on sale. Brand loyalty is, of course, a matter of degree. Some consumers will not switch for a moderate discount, but would switch for a large one or will occasionally buy another brand for convenience or variety.

The product of the marketing mix refers to the goods and / or services that the organisation will offer to the consumer. An organisation can achieve this by either creating a new food product, or by modifying or improving an existing food product.

Price

In profitably pricing the food, the manufacturer must keep in mind that the retailer adds approximately 50 percent to the price of a wholesale product. For example, a frozen food sold in a retail store for 250 INR generates an income of 100 INR for the manufacturer. This money has to pay for the cost of producing, packaging, shipping, storing, and selling the product.

Price encompasses the amount of money paid by the consumer in order to purchase the food product. When pricing the food products, the manufacturer must bear in mind that the retailer will add a particular percentage to the price on the wholesale product. This percentage amount differs globally.

The percentage is used to pay for the cost of producing, packaging, shipping, storing and selling the food product.

Promotion

Promoting a food to consumers is done out of store, in store, and on package. Advertisements on television and in magazines are attempts to persuade consumers to think favourably about a product, so that they go to the store to purchase the product. In addition to advertising, promotions can also include Sunday newspaper ads that offer coupons.

Promotion of the marketing mix is defined as the actions used to communicate a food product's features and benefits; therefore, persuading the consumer to purchase the product.

There are multiple avenues used to promote a food product to consumers.

- o Out-of-store
- o In-store
- o On packaging.

Food advertisements could be on
o TV
o Radio
o Print media
  • Social media
  • Billboards
  • Buses
  • Trains
  • Trams
  • Stations
  • Others

Additionally, promotions in magazines and newspapers may offer coupons for food products.

Place

Place refers to the distribution and warehousing efforts necessary to move a food from the manufacturer to a location where a consumer can buy it. It can also refer to where the product is located in a retail outlet (e.g. the end of an aisle; the top, bottom, or middle shelf; in a special display case, etc.).

Consumer focus helps marketers anticipate the demands of consumers, and production focus helps them respond to changes in the market. The result is a system that meets and influences the ever-changing demands of consumers.

Place refers to the activities that organisations go about in order to make its food product available to its consumers. This encompasses the distribution necessary to move a food product from the manufacturer to a location where it can be purchased by the consumer. Product location in a store is also a definition of place in the marketing mix.

Success of Food Marketing

Marketing success of a food product is determined by the fact as to how well the marketers knew the food product and the intended consumer.

In order to market its food products, an organisation must first understand whether its product will satisfy the consumer's needs better than competitors. In order to achieve this, an organisation must understand the four types of segmentation.

Geographic

An organisation must understand where it is marketing its food products in a geographical sense. Clarifying this will help an organisation to grasp which food products will satisfy the needs of a particular consumer culture.

Demographic

A food organisation must understand the demographic segment that it will be marketing towards. Points to keep in mind are a consumer's age, gender, education, social class, income, religion and ethnicity. All of these aspects can impact whether the consumer will prefer one food product over another.

Psychographic

A food organisation must understand its consumer psychographic profile. Factors such as lifestyle, personalities, opinions, activities and interests of its potential consumers must be considered. Identifying these aspects can help an organisation to improve its food products.

Behavior

A food organisation must understand how its consumers may behave towards a food product.

Considerations with Food Marketing

There are no set worldwide food marketing laws or legislation, therefore countries have the option to adopt their own legislation with regards to food marketing standards. (Restrict Food Marketing, 2016).

These standards are usually based on the values, culture and ethics of the particular country. Many countries worldwide have created laws in order to limit food marketing towards children with the goal of reducing rising obesity levels.

Obesity has been proven to have increased significantly with a link to the increase of food marketing and advertising in society.

Following a 'call out' from the World Health Organisation in 2006, many countries adopted to change marketing laws and legislation in order to protect children from persuasive advertisements directly targeted at them. These advertisements are strategically designed with special techniques in order to attract the children's attention.

Advertising to children at a young age is a well-established food marketing technique designed to encourage brand preference and holds many ethical dilemmas. Previous studies have concluded that children can recognize and mentally picture brand logos at the age of just six months old, and will verbally request brands at the age of 3 years.

Marketing Professor James McNeal acknowledged "The Drool Factor"—a study which recognized the fact that babies naturally stare down at their bibs while drooling to see where their drool lands. Customizing baby bibs with brand logos has become an effective way for food marketers to imprint their brand into the child's lifestyle, targeting them at a vulnerable young age resulting in brand recognition from the child. As a result, when the child is older they will continue to reciprocate warm, fond feelings towards the brand when encountering it in society.

In a nutshell the following questions and their discussions determine the advertising and marketing strategy for a food product.

1. What is the product?
2. Who is the market?
3. Developing a marketing strategy.
4. Where will it be sold?
5. What you be the charge for it?
6. How will potential customers be convinced to choose the product?

Chapter 19

Food Laws

The food processing industry, one of the largest industries in India, is widely recognized as an upcoming industry in India having huge potential for uplifting the agricultural economy, creation of large-scale processed food manufacturing and food chain facilities, and the resultant generation of employment and export earnings.

The Indian food processing industry is regulated by several laws which govern the aspects of sanitation, licensing and other necessary permits that are required to start up and run a food business. The legislation that dealt with food safety in India was the Prevention of Food Adulteration Act, 1954 (hereinafter referred to as "**PFA**"). The PFA had been in place for over five decades and there was a need for change due to varied reasons which include the changing requirements of our food industry.

The act brought into force in place of the PFA is the Food Safety and Standards Act, 2006 (hereinafter referred to as "**FSSA**") that overrides all other food related laws. It specifically repealed eight laws which were in operation prior to the enforcement of FSSA:

o The prevention of Food Adulteration Act, 1954

o The fruit products order, 1955

o The meat food products order, 1973

o The vegetable oil products (control) order, 1947

o The edible oils packaging (regulation) order, 1998

- The solvent extracted oil, de oiled meal, and edible flour (control) order, 1967
- The milk and milk products order, 1992
- Essential Commodities Act, 1955 (in relation to food)

NEED FOR THE NEW ACT

FSSA initiates harmonization of India's food regulations as per international standards. It establishes a new national regulatory body, the Food Safety and Standards Authority of India (hereinafter referred to as "FSSAI"), to develop science based standards for food and to regulate and monitor the manufacture, processing, storage, distribution, sale and import of food so as to ensure the availability of safe and wholesome food for human consumption. All food imports will therefore be subject to the provisions of the FSSA and rules and regulations which as notified by the Government on 5th of August, 2011 will be applicable.

KEY REGULATIONS OF FSSA

A. Packaging and Labelling

FSSA provides for separate packaging and labelling regula-tions known as Food Safety and Standards (packaging and labelling) Regulations, 2011 (hereinafter referred to as the **"Packaging and Labelling Regulations"**) which lay down the statutory and regulatory requirements for packaging and labelling of products. A plain reading of the Packaging and Labelling Regulations shows that there are different kinds of products: Pre-packaged, Proprietary and other specific products as mentioned in the regulations.

Regulation 2.12 of the Food Safety and Standards (Food products standards and food additives) Regulations, 2011 defines *proprietary food* as food that has not been standardized under these regulations. Regulation 1 (8) of the packaging and labelling regulations defines *prepackaged* or *pre-packed food*, as food, which is placed in a package of any nature, in such a manner that the contents cannot be changed without tampering it and which is ready for sale to the consumer.

The packaging and labelling regulations provide the general requirements for labelling of food products prescribed under the FSSA, as follows:

i. The particulars of declaration required under these Regulations to be specified on the label shall be in English or Hindi in Devnagri script: Provided that nothing herein contained shall prevent the use of any other language in addition to the language required under this regulation.

ii. Pre-packaged food shall not be described or presented on any label or in any manner that is false, misleading or deceptive or is likely to create an erroneous impression regarding its character in any respect;

iii. Label in pre-packaged foods shall be applied in such a manner that they will not become separated from the container;

iv. Contents on the label shall be clear, prominent, indelible and readily legible by the consumer under normal conditions of purchase and use;

v. Where the container is covered by a wrapper, the wrapper shall carry the necessary information or the label on the container shall be readily legible through the outer wrapper and not obscured by it.

In addition to these general requirements specified above, every package of food shall also carry the following information on the label: (i) name of the food; (ii) list of ingredients; (iii) nutritional information; (iv) declaration regarding veg. and non-veg; (v) declaration regarding food additives; (vi) name and complete address of the manufacturer; (vii) net quantity; (viii) lot/code/batch identification; (ix) date of manufacturing or packing; (x) best before and use by date; (xi) country of origin for imported food; and (xii) instructions for use.

Since a large variety of food products are being imported into India, under the Packaging and Labelling Regulations, it becomes necessary to mention the country of origin of the food on the label of food imported into India, and when a food undergoes processing in a second country which changes its nature, the country in which the processing is performed shall be considered to be the country of origin for the purposes of labelling.

Therefore, the above are the statutory and regulatory requirements that are to be complied with regard to labelling of products that are sold in the Indian market as "pre-packaged goods".

B. Signage and Customer Notices

Having briefly dealt with the statutory and regulatory requirements with respect to labelling of products, it is necessary to understand the statutory and regulatory requirements with respect to signage and customer notices more from the point of view of a food outlet. It is important to note that though the provisions of FSSA do not specifically provide for any statutory and regulatory requirements either for signage or customer notices, but it has certain provisions with regard to advertisement of products by food business operators.

Section 3 (1) (b) of FSSA defines the term *advertisement* (which includes a "notice") as any audio or visual publicity, representation or pronouncement made by means of any light, sound, smoke, gas, print, electronic media, internet or website and includes through any notice, circular, label, wrapper, invoice or other documents.

Section 24 of the FSSA provides that no advertisement shall be made of any food which is misleading or deceiving or contravenes the provisions, rules and regulations made there under. No person shall engage himself in any unfair trade practice for purpose of promoting the sale, supply, use and consumption of articles of food or adopt any unfair or deceptive practice including the practice of making any statement, whether orally or in writing or by visible representation which:

    i. falsely represents that the foods are of a particular standard, quality, quantity or grade-composition;

    ii. makes a false or misleading representation concerning the need for, or the usefulness;

    iii. gives to the public any guarantee of the efficacy that is not based on an adequate or scientific justification thereof, provided that where a defence is raised to the effect that such guarantee is based on adequate or scientific justification, the burden of proof of such defence shall lie on the person raising such defence.

FSSA being applicable to all food business operators in India, the provision with regard to advertisements would have to be complied with.

C. Licensing Registration and Health And Sanitary Permits

It is also important to note that FSSA, being the only legislation applicable to the food industry throughout the country, will also apply as far as the national health and sanitary permits are concerned.

The Food Safety and Standards (Licensing and Registration of Food Business) Regulations, 2011 (hereinafter referred to as "**License and Registration Regulations**") govern the aspect of license and registration of a food business operator.

Under Regulation 2.1 of the License and Registration Regulations, all food business operators in the country are required to be registered or licensed in accordance with the License and Registration Regulations, hence no person shall commence any food business unless a valid license is possessed by the food business operator, and the conditions with regard to safety, sanitary and hygienic requirements have to be complied with at all times by them.

One of the prime purposes of these conditions is to ensure that the food business operator maintains sanitary and hygienic standards as specified in each food category. It is hereby recognized and declared as a matter of legislative determination that in the field of human nutrition, safe, clean, wholesome food is indispensable to the health and welfare of the consumer of the country.

It shall be deemed the responsibility of the food business to comply with the labelling, safety and health and sanitary requirements laid down in the License and Registration Regulations. The labelling requirements are specified under the regulations and they need to be complied with at all times especially with regard to pre-packaged goods.

Penalties

The FSSA provides for penalties in case of any non-compliance. Generally, non-compliance with various provisions of the FSSA may attract penalty of up to Two Lakh Rupees (*approx USD*

4000). However, under Section 63, it provides that if any person or food business operator (except the persons exempted from licensing under Sub-section (2) of Section 31 of FSSA), himself or by any person on his behalf who is required to obtain license, manufacturers, sells, stores or distributes or imports any article of food without license, shall be punishable with imprisonment for a term which may extend to six months and also with a fine which may extend to Five Lakh Rupees (*approx USD 9000*).

Other Licenses

The FSSA being a central act has to be complied with by all the food business operators in the country. However, India being a big market, each state may have local laws which may also need to be complied with. Some of the other approvals and licenses that a food operator may be required to obtain from various authorities under other laws include: Health and trade licenses from the municipal corporation of the relevant area, environmental clearance, no-objection certificate for fire prevention and safety, registration under the Police Act of the respective city/state, verification certificate under the Standards of Weights and Measures Act, 1976 for each of the outlets issued by the Department of Legal Metrology of the respective areas, registration under the Shops and Establishments Act of the respective state, eating house license and liquor license.

A license for playing music in restaurants is also required for playing recorded or live music. It is mandatory for a food business to obtain insurance from any insurance company with regard to public policy, product liability, fire policy, building and assets. Other insurances though are not mandatory may be useful if taken.

Some of the other registrations and permissions may include registration under the Employees' Provident Funds and Miscellaneous Provisions Act, 1952 if it is engaging more than 20 employees. Registration is also required under the Central Excise Act, 1944 as in respect of goods specified in Third Schedule of the said act, repacking, re-labelling, putting or altering retail sale price, etc. will fall into the category of manufacture. Subject to applicability, other statutory and regulatory compliances may also include registrations under

Income Tax Act, 1861, Customs Act, 1962, sales tax, service tax and other labor laws.

Foreign Direct Investment in the Food Processing Industry

Foreign Direct Investment (hereinafter referred to as "FDI") is permissible for all the processed food products under 100% automatic route (except for items reserved for micro, small and medium enterprises, where FDI is permissible under automatic route up to 24%), subject to applicable laws/regulations/ securities and other conditions.

Food Standards

ISI Standard

When it comes to quality assurance, no other standard does that better than the ISI mark. This mark of quality specifies that the product with this mark can be consumed with surety. The committee formed for evaluation of quality of various food products comprise industrialists, government representatives, consumer representations to get a broad outlook of the needs and expectations about the product. These committee then formulate the Indian Standards Institution (ISI). Standards are laid for various products like fruits and vegetables, animal products, plant products, spices and condiments as well as variety of processed foods.

The ISI mark is given to the products that conform to these standards laid down by the committee. Once the product gets the ISI seal, it gains a respect among consumers and gives rise to a trust for the product.

The quality of the products with the ISI seal is periodically checked for quality and evaluated in the laboratories maintained by ISI in many cities like Delhi, Mumbai, Chennai, Kolkata and Chandigarh.

AGMARK

The standard is an abbreviation of Agricultural Marking.

Under the Agricultural Produce Act of 1937, the Directorate of Marketing and Inspection of the Government of India set up the AGMARK standard. As the ISI seal is synonymous with purity and quality so is the AGMARK seal. The products with the mark are considered to be of good quality and pure.

The seal is given after careful tests and deliberations. The various physical and chemical characteristics of the product are taken into account while giving the AGMARK seal. The food commodities which come under the Act are cereals (and products), legumes (and products), eggs, butter, ghee, oils and oil seeds. The products are categorised into grades depending on the qualities and purity of each sample. Generally, four grades are given to food commodities:

1. Special
2. Good
3. Fair
4. Ordinary

The product with grade 1 is the best and the rest follow. These grades keep in mind the colour, texture, feel, moisture, weight, size, variety, fat content and nutritional value besides other factors while testing the product.

To keep a check on the products the Directorate of Marketing and Inspection works with the help of 21 laboratories and 50 suboffices all over India. The main Central AGMARK Laboratory is at Nagpur and foresees the work in research and development.

People with adequate experience and standing in the market are granted "The Certificate of Authorisation". This entails them with powers and authorisation to work in the field. The prior procedures, before the AGMARK label is given to a product like selection, processing, grading and packing of food commodities, are conducted with the staff of the Directorate of Marketing and Inspection or staff of the State Government in attendance.

Grading some commodities like ghee, vegetable oils, flour, spices, butter and honey is compulsory while tobacco, walnuts, basmati rice, essential oils, onions, potatoes, if meant for export, are graded. The importers of these products are assured of the quality by the AGMARK seal. The products with the AGMARK seal are costlier than others due to added costs of processing and grading procedures. But due to the benefits of the programme the added cost is no hindrance.

1. The consumers feel assured and trust the quality of the product.
2. The grade serves to create a common and standard language to explain the quality of the product facilitating trade and commerce. It prevents the hassles associated with frequent quality checks.
3. The producers are saved from being exploited as they can demand a price for their product and can bargain effectively.

Export Inspection Council

The council keeps a check on articles meant to be exported. All food commodities meant to be sent to other countries for trade are screened by this body. Articles below a certain standard are rejected and the consignment not exported. This is done to keep the goodwill alive and the trust in Indian products safe. If a spoilt article is exported it will give a bad name to the country and close possible routes for trade. Foods as canned foods (fruit juices), frozen food (meat products) are scrutinised by the council and then given a final nod.

Chapter 20

Post-harvest Losses and Technology

Post-harvest handling is the stage of crop production immediately following harvest, including cooling, cleaning, sorting and packing. The instant a crop is removed from the ground, or separated from its parent plant, it begins to deteriorate. Post-harvest treatment largely determines final quality, whether a crop is sold for fresh consumption, or used as an ingredient in a processed food product.

Harvest and post-harvest loss of India's major agricultural produce is estimated at Rs 92,651 crores, according to data published by the Ministry of Food Processing Industries on 9 August, 2016.

About 16 percent of fruits of vegetables, valued at Rs 40,811 crores, were lost, according to an analysis of production data between 2012 and 2014, at wholesale prices, by the Central Institute of Post-Harvest Engineering and Technology, Ludhiana (Punjab).

The food processing ministry also reported that 7 percent of meat, valued at Rs 3,942 crores, was lost, about 60 percent during storage.

Almost 7,000 cold stores were created under four central programmes between 2007 and 2014 to curb supply chain losses. These projects cost Rs 2,395 crores. An additional 609 projects have been sanctioned and subsidy of Rs 660 crores provided between 2014 and 2016.

Elimination of post-harvest losses is a critical component to safeguard future food security. Decline in post-harvest losses will enhance global food security, an emergent concern with intensifying food prices and increased weather variability.

OBJECTIVES OF POST-HARVEST HANDLING

The most important goals of post-harvest handling are keeping the product cool, to avoid moisture loss and slow down undesirable chemical changes, and avoiding physical damage such as bruising, to delay spoilage. Sanitation is also an important factor, to reduce the possibility of pathogens that could be carried by fresh produce, for example, as residue from contaminated washing water. After the field, post-harvest processing is usually continued in a packing house.

Initial post-harvest storage conditions are critical to maintaining quality. Each crop has an optimum range of storage temperature and humidity. Also, certain crops cannot be effectively stored together, as unwanted chemical interactions can result.

Various methods of high-speed cooling, and sophisticated refrigerated and atmosphere-controlled environments, are employed to prolong freshness, particularly in large-scale operations.

Regardless of the scale of harvest, from domestic garden to industrialised farm, the basic principles of post-harvest handling for most crops are the same: Handle with care to avoid damage (cutting, crushing, bruising), cool immediately and maintain in cool conditions, and cull (remove damaged items).

Once harvested, vegetable and fruit are subject to the active process of senescence. Numerous biochemical processes continuously change the original composition of the crop until it becomes unmarketable. The period during which consumption is considered acceptable is defined as the time of "post-harvest shelf life".

Post-harvest shelf life is typically determined by objective methods that determine the overall appearance, taste, flavour, and texture of the commodity. These methods usually include a combination of sensorial, biochemical, mechanical, and colourimetric (optical) measurements. A recent study attempted (and failed) to discover a biochemical marker and fingerprint methods as indices for freshness.

Causes of Post-harvest Losses in Fruits and Vegetables

Fruits and vegetables are living parts of plant and contain 65 to 95 percent water. When food and water reserves are exhausted, produce dies and decays. Anything that increases the rate at which a product's food and water reserves are used up increases the likelihood of losses. Increases in normal physiological changes can be caused by high temperature, low atmospheric humidity and physical injury. Such injury often results from careless handling, causing internal bruising, splitting and skin breaks, thus rapidly increasing water loss.

Respiration is a continuing process in a plant and cannot be stopped without damage to the growing plant or harvested produce. It uses stored starch or sugar and stops when reserves of these are exhausted, leading to ageing. Respiration depends on a good air supply. When the air supply is restricted fermentation instead of respiration can occur. Poor ventilation of produce also leads to the accumulation of carbon dioxide. When the concentration of carbon dioxide increases, it will quickly ruin produce.

Fresh produce continues to lose water after harvest. Water loss causes shrinkage and loss of weight. The rate at which water is lost varies according to the product. Leafy vegetables lose water quickly because they have a thin skin with many pores. Potatoes, on the other hand, have a thick skin with a few pores. But whatever the product, to extend shelf or storage life the rate of water loss must be minimal. The most significant factor is the ratio of the surface area of the fruit or vegetable to its volume. The greater the ratio the more rapid will be the loss of water. The rate of loss is related to the difference between the water vapour pressure inside the produce and in the air. Produce must therefore be kept in a moist atmosphere.

Diseases caused by fungi and bacteria cause losses but virus diseases, common in growing crops, are not a major post-harvest problem. Deep penetration of decay makes infected produce unusable. This is often the result of infection of the produce in the field before harvest. Quality loss occurs when the disease affects only the surface. Skin blemishes may lower the sale price but do not render a fruit or vegetable inedible. Fungal and bacterial diseases are spread by microscopic spores, which are distributed in the air and soil and via decaying plant material. Infection after harvest can occur at any time. It is usually the result of harvesting or handling injuries.

Ripening occurs when a fruit is mature. Ripeness is followed by senescence and breakdown of the fruit. The category "fruit" refers also to products such as aubergine, sweet pepper and tomato.

Non-climacteric fruit only ripen while still attached to the parent plant. Their eating quality suffers if they are harvested before fully ripe as their sugar and acid content does not increase further. Examples are citrus, grapes and pineapple.

Early harvesting is often carried out for export shipments to minimise loss during transport, but a consequence of this is that the flavour suffers.

Climacteric fruit are those that can be harvested when mature but before ripening has begun. These include banana, melon, papaya, and tomato. In commercial fruit marketing the rate of ripening is controlled artificially, thus enabling transport and distribution to be carefully planned.

Ethylene gas is produced in most plant tissues and is important in starting off the ripening process. It can be used commercially for the ripening of climacteric fruits. However, natural ethylene produced by fruits can lead to in-storage losses. For example, ethylene destroys the green colour of plants. Leafy vegetables will be damaged if stored with ripening fruit. Ethylene production is increased when fruits are injured or decaying and this can cause early ripening of climacteric fruit during transport.

Produce can be damaged when exposed to extremes of temperature. Levels of tolerance to low temperatures are important when cool storage is envisaged. All produce will freeze at temperatures between 0 and –2 degrees Celsius. Although a few commodities are tolerant of slight freezing, bad temperature control in storage can lead to significant losses.

Some fruits and vegetables are also susceptible to contaminants introduced after harvest by use of contaminated field boxes; dirty water used for washing produce before packing; decaying, rejected produce lying around packing houses; and unhealthy produce contaminating healthy produce in the same packages.

Losses directly attributed to transport can be high, particularly in developing countries. Damage occurs as a result of careless handling of packed produce during loading and unloading; vibration (shaking) of the vehicle, especially on bad roads; and poor stowage, with packages often squeezed into the vehicle in order to maximise revenue for the transporters. Overheating leads to decay, and increases the rate of water loss. In transport it can result from using closed vehicles with no ventilation; stacking patterns that block the movement of air; and using vehicles that provide no protection from the sun. Breakdowns of vehicles can be a significant cause of losses in some countries, as perishable produce can be left exposed to the sun for a day or more while repairs are carried out.

At the retail marketing stage losses can be significant, particularly in poorer countries. Poor-quality markets often provide a little protection for the produce against the elements, leading to rapid produce deterioration. Sorting of produce to separate the saleable from the unsaleable can result in high percentages being discarded, and there can be high weight loss from the trimming of leafy vegetables. Arrival of fresh supplies in a market may lead to some existing, older stock being discarded, or sold at very low prices.

Causes of Post-harvest Losses in Grains

Grains may be lost in the pre-harvest, harvest and post-harvest stages. Pre-harvest losses occur before the process of harvesting begins, and may be due to insects, weeds and rusts. Harvest losses occur between the beginning and completion of harvesting, and are primarily caused by losses due to shattering.

Post-harvest losses occur between harvest and the moment of human consumption. They include on-farm losses, such as when grain is threshed, winnowed and dried, as well as losses along the chain during transportation, storage and processing. Important in many developing countries, particularly in Africa, are on-farm losses during storage, when the grain is being stored for auto-consumption or while the farmer awaits a selling opportunity or a rise in prices.

The main cause of loss during drying is the cracking of grain kernels that are eaten whole, such as rice. Some grains may also be lost during the drying process. However, failure to dry crops adequately can lead to much higher levels of loss than poor-quality drying, and may result in the entire harvest becoming inedible. Adequate drying by farmers is essential if grains are to be stored on-farm and poorly dried grains for the market need to be sold quickly to enable the marketing-processing chain to carry out adequate drying before the grains become spoilt. With high moisture content, grain is susceptible to mould, heating, discolouration and a variety of chemical changes. Ideally, most grains should be dried to acceptable levels within 2–3 days of harvest.

One of the problems in assessing levels of post-harvest loss is in separating weight loss caused by the very necessary drying operations from weight loss caused by other controllable factors.

Milling to remove the outer coats from a grain may take place in one or more stages. For paddy rice considerable mechanical effort is needed to remove these layers. Any weakness in the kernel will be apparent at this stage. Even with grain in perfect condition, correctly set milling and polishing machinery is essential to yield high processing outturns. Complete separation of edible from less-desired products is always difficult to achieve but, even so, there are significant differences in milling efficiency. In the case of rice, milling outturns can vary from 60% or less to around 67%, depending on the efficiency of the mill. Even a 1% increase in yield of whole grain rice can thus result in huge increases in national food resources.

Contamination by moulds is mainly determined by the temperature of the grain and the availability of water and oxygen. Moulds can grow over a wide range of temperatures, but the rate of growth is lower with lower temperature and less water availability. The interaction between moisture and temperature is important. Maize, for example, can be stored for one year at a moisture level of 15% and a temperature of 15°C. However, the same maize stored at 30°C will be substantially damaged by moulds within three months. Insects and mites (arthropods) can, of course, make a significant contribution towards the deterioration of grain, through the physical damage and nutrient losses caused by their activity.

Importance of Post-harvest Technology

Due to protection, conservation, processing, packaging, distribution, marketing, and utilization to meet the food and nutritional standards of the people in relation to their needs, post-harvest technology had been developed in consonance with the needs of each society to stimulate agricultural production, prevent post-harvest losses, improve efficiency and add value to the products. During this process, proper uses of post-harvest technology would generate employment, reduce poverty and stimulate growth of related economic sectors.

Importance of postharvest technology lies in the fact that it has capability to meet food requirement of growing world population by eliminating avoidable losses making more nutritive food items with higher values by proper processing, storage, packaging, transport and marketing. Use of appropriate post-harvest technology not only reduces the post-harvest and storage losses or adds value to the product, but more importantly it provides the potential of higher employment including fortification of agricultural and agro-industries. And at the end, establish food security for green and happiness society. Innovation Center was established to serve the purpose through an inter-disciplinary and multi-dimensional approach, which included, research capability, scientific creativity, technological innovations, productive sectors participation, human resources development, all of which must respond in an integrated manner to the developmental needs of the country.

Interventions to Minimize or Avoid Losses

There is a wide range of post-harvest technologies that can be adopted to improve losses throughout the process of pre-harvest, harvest, cooling, temporary storage, transport, handling and market distribution. Recommended technologies vary depending on the type of loss experienced.

However, all interventions must meet the principle of cost-effectiveness. In theory it should be possible to reduce losses substantially but in practice this may be prohibitively expensive. Especially for small farms, for which it is essential to reduce losses, it is difficult to afford expensive and work-intensive technologies.

The various methods used are:
1. Physical treatments
2. Storage condition considerations
3. Biochemical products
    a. Microbial antagonists
    b. Plant derived products
    c. Animal derived products

1. Physical Treatments

a. ***Cooling methods and temperatures*:** Several methods of cooling are applied to produce after harvesting to extend shelf life and maintain a fresh-like quality. Some of the low temperature treatments are unsuitable for simple rural or village treatment but are included for consideration as follows:

i. *Precooling*: Fruit is precooled when its temperature is reduced from 3 to 6°C (5 to 10°F) and is cool enough for safe transport. Precooling may be done with cold air, cold water (hydrocooling), direct contact with ice, or by evaporation of water from the product under a partial vacuum (vacuum cooling). A combination of cooled air and water in the form of a mist called hydro aircooling is an innovation in cooling of vegetables.

ii. *Air precooling*: Precooling of fruits with cold air is the most common practice. It can be done in refrigerator, cars, storage rooms, tunnels, or forced air-coolers (air is forced to pass through the container via baffles and pressure differences).

iii. *Icing*: Ice is commonly added to boxes of produce by placing a layer of crushed ice directly on the top of the crop. An ice slurry can be applied in the following proportion: 60% finely crushed ice, 40% water, and 0.1% sodium chloride to lower the melting point. The water to ice ratio may vary from 1:1 to 1:4.

iv. *Room cooling*: This method involves placing the crop in cold storage. The type of room used may vary, but generally consists of a refrigeration unit in which cold air is passed through a fan. The circulation may be such that air is blown across the top of the room and falls through the crop by convection. The main advantage is cost because no specific facility is required.

v. *Forced air-cooling*: The principle behind this type of precooling is to place the crop into a room where cold air is directed through the crop after flowing over various refrigerated metal coils or pipes. Forced air-cooling systems blow air at a high velocity leading to desiccation of the crop. To minimize this effect, various methods of humidifying the cooling air have been designed such as blowing the air through cold water sprays.

vi. *Hydrocooling*: The transmission of heat from a solid to a liquid is faster than the transmission of heat from a solid to a gas. Therefore, cooling of crops with cooled water can occur quickly and results in zero loss of weight. To achieve high performance, the crop is submerged in cold water, which is constantly circulated through a heat exchanger. When crops are transported around the packhouse in water, the transport can incorporate a hydrocooler. This system has the advantage wherein the speed of the conveyer can be adjusted to the time required to cool the produce. Hydrocooling has a further advantage over other precooling methods in that it can help clean the produce. Chlorinated water can be used to avoid spoilage of the crop. Hydrocooling is commonly used for vegetables, such as asparagus, celery, sweet corn, radishes, and carrots, but it is seldom used for fruits.

vii. *Vacuum cooling*: Cooling in this case is achieved with the latent heat of vaporization rather than conduction.

At normal air pressure (760 mmHg) water will boil at 100°C. As air pressure is reduced so is the boiling point of water, and at 4.6 mmHg water boils at 0°C. For every 5 or 6°C reduction in temperature, under these conditions, the crop loses about 1% of its weight (Barger, 1961). This weight loss may be minimized by spraying the produce with water either before enclosing it in the vacuum chamber or towards the end of the vacuum cooling operation (hydrovacuum cooling). The speed and effectiveness of cooling is related to the ratio between the mass of the crop and its surface area. This method is particularly suitable for leaf crops such as lettuce. Crops like tomatoes having a relatively thick wax cuticle are not suitable for vacuum cooling.

b. *High temperatures*: Exposure of fruits and vegetables to high temperatures during post-harvest reduces their storage or marketable life. This is because as living material, their metabolic rate is normally higher with higher temperatures. High temperature treatments are beneficial in curing root crops, drying bulb crops, and controlling diseases and pests in some fruits. Many fruits are exposed to high temperatures in combination with ethylene (or another suitable gas) to initiate or improve ripening or skin colour.

2. Storage Condition Considerations

The marketable life of most fresh vegetables can be extended by prompt storage in an environment that maintains product quality. The desired environment can be obtained in facilities where temperature, air circulation, relative humidity, and sometimes atmosphere composition can be controlled. Storage rooms can be grouped accordingly as those requiring refrigeration and those that do not. Storage rooms and methods not requiring refrigeration include: *In situ*, sand, coir, pits, clamps, windbreaks, cellars, barns, evaporative cooling, and night ventilation.

a. **In situ**: This method of storing fruits and vegetables involves delaying the harvest until the crop is required. It can be used in some cases with root crops, such as cassava, but means that the land on which the crop was grown will remain occupied and a new crop cannot be planted. In colder climates, the crop may be exposed to freezing and chilling injury.

b. **Sand or coir**: This storage technique is used in countries like India to store potatoes for longer periods of time, which involves covering the commodity under ground with sand.

c. **Pits** or trenches are dug at the edges of the field where the crop has been grown. Usually pits are placed at the highest point in the field, especially in regions of high rainfall. The pit or trench is lined with straw or other organic material and filled with the crop being stored, then covered with a layer of organic material followed by a layer of soil. Holes are created with straw at the top to allow for air ventilation, as lack of ventilation may cause problems with rotting of the crop.

d. **Clamps**: This has been a traditional method for storing potatoes in some parts of the world, such as Great Britain. A common design uses an area of land at the side of the field. The width of the clamp is about 1 to 2.5 m. The dimensions are marked out and the potatoes piled on the ground in an elongated conical heap. Sometimes straw is laid on the soil before the potatoes. The central height of the heap depends on its angle of repose, which is about one-third the width of the clump. At the top, straw is bent over the ridge so that rain will tend to run off the structure. Straw thickness should be from 15 to 25 cm when compressed. After two weeks, the clamp is covered with soil to a depth of 15–20 cm, but this may vary depending on the climate.

e. **Windbreaks** are constructed by driving wooden stakes into the ground in two parallel rows about 1 m apart. A wooden platform is built between the stakes about 30 cm from the ground, often made from wooden boxes. Chicken wire is affixed between the stakes and across both ends of the windbreak. This method is used in Britain to store onions (Thompson, 1996).

f. **Cellars**: These underground or partly underground rooms are often beneath a house. This location has good insulation, providing cooling in warm ambient conditions and protection from excessively low temperatures in cold climates. Cellars have traditionally been used at domestic scale in Britain to store apples, cabbages, onions, and potatoes during winter.

g. **Barns.** A barn is a farm building for sheltering, processing, and storing agricultural products, animals, and implements. Although there is no precise scale or measure for the type or size of the building, the term barn is usually reserved for the largest or most important structure on any particular farm. Smaller or minor agricultural buildings are often labelled sheds or outbuildings and are normally used to house smaller implements or activities.

h. **Evaporative cooling.** When water evaporates from the liquid phase into the vapour phase energy is required. This principle can be used to cool stores by first passing the air introduced into the storage room through a pad of water. The degree of cooling depends on the original humidity of the air and the efficiency of the evaporating surface. If the ambient air has low humidity and is humidified to around 100% RH, then a large reduction in temperature will be achieved. This can provide cool moist conditions during storage.

i. **Night ventilation.** In hot climates, the variation between day and night temperatures can be used to keep stores cool. The storage room should be well insulated when the crop is placed inside. A fan is built into the store room, which is switched on when the outside temperature at night becomes lower than the temperature within. The fan switches off when the temperatures equalize. The fan is controlled by a differential thermostat, which constantly compares the outside air temperature with the internal storage temperature. This method is used to store bulk onions.

j. **Controlled atmospheres** are made of gastight chambers with insulated walls, ceiling, and floor. They are increasingly common for fruit storage at larger scale. Depending on the species and variety, various blends of O_2, CO_2, and N_2 are required. Low content O_2 atmospheres

(0.8 to 1.5%), called ULO (ultra-low oxygen) atmospheres, are used for fruits with long storage lives (e.g. apples).

3. Bio-pesticides

Globally, food safety has emerged as a key element in decay control programs. Understandably, alternatives to chemical pesticides or products that allow reduced usage in terms of fewer or reduced rates of application are beginning to appear in the market. A part of the solution to this problem is controlled through the use of biopesticides, the foundation on which sustainable, non-polluting pest control for tomorrow's farms must be built.

There is no formally agreed definition of a biopesticide. Biologicals or biopesticides are certain types of mass-produced agent derived from a living micro-organism or a natural product and sold for the control of plant pests. In a much simpler way we can say that these are pest management tools that are based on beneficial microorganisms (bacteria, viruses, fungi and protozoa), beneficial nematodes or other safe, biologically based active ingredients. The use of these materials is widespread with applications to foliage, turf, soil, or other environments of the target insect pests.

a. **Microbial pesticides:** Naturally occurring or genetically controlled bacteria, fungi, viruses and protozoa are all being used for the biological control of pestiferous insects, plant pathogens, weeds, microorganism (e.g., a virus bacterium, fungus, nematode or protozoan) as the active ingredient. These pesticides can control many different kinds of pests, although each separate active ingredient is relatively specific for its target pests. Microbial pesticides need to be continuously monitored to ensure that they do not become capable of harming non-target organisms, including humans.

b. **Plant-derived products:** The use of locally available plants for the control of pests and pathogens is an age-old technology in many parts of the world. Farmers in their traditional wisdom have identified and used a variety of plant products and extracts for pest control, especially in storage. As many as 2121 plant species are reported to possess pest management properties, 1005 species of plants exhibiting insecticide properties, 384 with antifeedant properties, 297 with repellant properties, 27 with attractant properties and 31 with growth inhabiting properties have been identified.

The efficacy of essential oils and vegetable oils in preventing infestation of stored product pests such as bruchids, rice and maize weevils has been well documented. Root extracts of *Tagetes* or *Asparagus* as nematicide and *Chenopodium* and *Bougainvillea* as antivirus have also been reported.

Higher plants are the source of a wide spectrum of secondary metabolites such as alkaloids, essential oils, flavonoids, phenolics, quinines, saponins, sterols and tannins, which offer resistance to pathogens some of these can be used as biopesticides. They include, for example, pyrethrins, which are fast-acting insecticidal compounds produced by *Chrysanthemum cinerariaefolium*.

Abamectin, a macrocyclic lactone compound, produced by *Streptomyces avermitilis* is active against a range of pest species but resistance has developed in tetranychid mites to it also.

c. Animal Derived Products

○ **Chitosan** is a linear polysaccharide composed of randomly distributed β-(1→4)-linked D-glucosamine (deacetylated unit) and N-acetyl-D-glucosamine (acetylated unit). It is made by treating the chitin shells of shrimp and other crustaceans with an alkaline substance, like sodium hydroxide.

Chitosan has a number of commercial and possible biomedical uses. It can be used in agriculture as a seed treatment and biopesticide, helping plants to fight off fungal infections. In winemaking, it can be used as a fining agent, also helping to prevent spoilage. In industry, it can be used in a self-healing polyurethane paint coating. In medicine, it may be useful in bandages to reduce bleeding and as an antibacterial agent; it can also be used to help deliver drugs through the skin.

More controversially, chitosan has been asserted to have use in limiting fat absorption, which would make it useful for dieting, but there is evidence against this.

Other uses of chitosan that have been researched include use as a soluble dietary fiber.

Pest Control and Decay

Crops may be immersed in hot water before storage or marketing to control disease. A common disease of fruits known as anthracnose, caused by the infection of fungus *Colletotrychum spp.* can be successfully controlled in this way. Combining appropriate doses of fungicides with hot water is often effective in controlling disease in fruits after harvesting.

Fruit and vegetable decay is also caused by storage conditions. Too low temperatures can cause injury during refrigeration of fruits and vegetables. High temperatures can cause softening of tissues and promote bacterial diseases. The damage that microorganisms inflict on fresh fruits and vegetables is mainly in the physical loss of edible matter, which may be partial or total.

Novel methods of post-harvest control

1. Use of yeast solution
2. Use of essential oils from citrus sources
3. Use of UV light source

Bibliography

Adams, Catherine (2007). "Reframing the Obesity Debate: McDonald's Role May Surprise You". Journal of Law, Medicine, and Ethics. 35: 154–157. doi:10.1111/j.1748-720X.2007.00120.x.

Allen, MW, Gupta, R and Monnier, A (2008). The interactive effect of Cultural Symbols and Human Values on Taste Evaluation. Journal of Consumer Research Vol. 35, 294–308.

Annual Report 2014 (PDF). Thai President Foods. 2015. Retrieved 9 Jun 2015. Appetite, 86 (2015), pp. 3–18.

Ares, G, and Gámbaro, A (2007). Influence of gender, age and motives underlying food choice on perceived healthiness and willingness to try functional foods. Appetite, 49(1), 148–158.

Arndt, Michael. "McDonald's 24/7." Business Week February 4, 2007.

Arya, SS, 1990, Grain based Snack and convenience foods. Indian food Packer, Sept–Oct, 17–34.

Askegaard, S and Madsen, TK (1998). The local and the global: exploring traits of homogeneity and heterogeneity in European food cultures. International Business Review, 7, 549–568.

Avantina Sharma (2017). "Textbook of Food Science and Technology", CBS Publishers and Distributors.

Barthomeuf, L, Rousset, S and Droit-Volet, S. (2009). Emotion and food. Do the emotions expressed on other people's faces affect the desire to eat liked and disliked food products? Appetite, 52, 27–33 .

Beech, Hannah (13 November 2006). "Momofuku Ando". Time.

Belitz, HD; Grosch, Werner; Schieberle, Peter (2009-01-15). Food Chemistry. Springer Science & Business Media. ISBN 9783540699330.

Bench Marking of Food Processing Technologies – NCOFTECH 2011. Indian Crop Processing Technology, Ministry of Food Processing Industries, Govt. Of India, Thanjavur, Tamil Nadu.

Bhide, Amar V. The Origin and Evolution of New Business: Oxford University Press, New York, 2000.

Bix, L; Nora Rifon; Hugh Lockhart; Javier de la Fuente (2003). *The Packaging Matrix: Linking Package Design Criteria to the Marketing Mix (PDF).* IDS Packaging. Archived from the original (PDF) on 2008-12-17. Retrieved December 11, 2008.

Brand Positioning: Strategies for Competitive Advantage—Subroto Sengupta. pp. 5–6.

CP Herman, DA Roth, J Polivy.

Cartere, JY (2009). *TV, Food Marketing and Childhood Obesity.*

Chandon, P, and Wansink, B (2012). Does food marketing need to make us fat? A review and solutions. *Nutrition Reviews, 70*(10), 571–593.

Consumer Behavior in Action: Real-Life Applications for Marketing Managers, Geoffrey P. Lantos. p. 45.

Cross-sectional examination of physical and social contexts of episodes of eating and drinking in a national sample of US adults.

Data Bank on Economic Parameters of the Food Processing Sector —2011 published by the Ministry of Food Processing Industries, Government of India, New Delhi (available at: http://mofpi.nic. in/images/File/DataBank_SectoralDatabaseFPIs_140212.pdf)

Edible Coatings to Improve Food Quality and Food Safety and Minimize Packaging Cost, Usda, 2011, retrieved 18 March 2013.

Effects of the presence of others on food intake: a normative interpretation.

e-Krishi Shiksha study material

EU Agricultural Markets *Briefs* No 6 | June 2015

FDA

Food and Eating in Medieval Europe. Martha Carlin and Joel T. Rosenthal (editors). The Hambledon Press, London. 1998. ISBN 1-85285-148-1

Food Safety Practices Guidance for Fresh Non-Filled Alimentary Paste Manufacturers Guidance Document Repository (GDR).

Food Science and Technology, *by* Avantika Sharma. International Book Distributing Co., Lucknow (ISBN: 81-8189-097-3).

Frechette, S. (2015). Food Marketing as a Relevant Determinant of Childhood Obesity: The Link Between Exposure to TV Food Advertising and Children's Body Weight. *Annals of Spiru Haret University, Journalism Studies, 16*(2), 25–31.

Freeman, B, Kelly, B, Baur, L, Chapman, K, Chapman, S, Gill, T, King, L (December 2014). "Digital Junk: Food and beverage marketing on Facebook". American Journal of Public Health.

Fu, Binxiao (2007). "Asian noodles: History, classification, raw materials, and processing". Food Research International. 41: 888–902. doi:10.1016/j. foodres.2007.11.007 – via Elsevier Science Direct.

Ghosh, Bishwanath entrepreneurship development in India: National Publishing House, Jaipur and New Delhi 2000.

Gupta and Srinivasan, Entrepreneurial Development: Sultan Chand and Sons, 1995.

Gupta CB and Khanka SS, Entrepreneurship and small business management: Sultan Chand and Sons, 2003.

Handbook of Processing and Utilization in Agriculture, By: Wolff, IA CRC series in Agriculture, Volume-II, Part-I. CRC Press, Florida (ISBN: 0-8493-3872-7).

Hans-Jürgen Bässler und Frank Lehmann: *Containment Technology: Progress in the Pharmaceutical and Food Processing Industry.* Springer, Berlin 2013, ISBN 978-3642392917

Heldman, DR ed (2003). "Encyclopedia of Agricultural, Food, and Biological Engineering". New York: Marcel Dekker.

Hellesvig-Gaskell, Karen. "Definition of Fast Foods LIVESTRONG. COM". LIVESTRONG.COM. Retrieved May 3, 2016.

Hogan, David. *Selling 'em by the Sack: White Castle and the Creation of American Food.* New York: New York University Press, 1997.

Hope Ngo (23 February 2001). "CNN.com – Instant noodles a health hazard: report – February 23, 2001". CNN. Retrieved 7 November 2012.

http://agricoop.nic.in/Agristatistics.htm (surfed on February 22, 2011).

http://mofpi.nic.in (surfed on February 24, 2011).

http://www.bbc.com/news/health-26254989

http://www.dinero.com/edicion-impresa/negocios/articulo/el-rey-fideos/182740

http://www.entrepreneurswebsite.com/2011/03/18/food-processing-industry-in-india/ (surfed on February 24, 2011).

http://www.portafolio.co/negocios/empresas/venimos-colombia-crear-mercado-nissin-foods-67850

IGNOU study material

International Journal of Management Research & Trends ISSN: 0976-9781 Volume 4, Issue 1, 2013 ©Kaur & Singh.

Kapuge, K. (2016). Determinants of Organic Food Buying Behavior: Special Reference to Organic Food Purchase Intention of Sri Lankan Customers. *Procedia Food Science*, 6, 303–308. doi:10.1016/j.profoo.2016.02.060

Kline, S. (2010). *Globesity, Food Marketing and Family Lifestyles.*

Kroc, Ray with Robert Anderson. *Grinding It Out: The Making of McDonald's.* St. Martin's Press, 1992.

Laudan, Rachel (September–October 2010). "In Praise of Fast Food". UTNE Reader. Retrieved 2010-09-24. Where modern food became available, people grew taller and stronger and lived longer.

Laudan, Rachel (September–October 2010). "In Praise of Fast Food". UTNE Reader. Retrieved 2010-09-24. If we fail to understand how scant and monotonous most traditional diets were, we can misunderstand the "ethnic foods" we encounter in cookbooks, at restaurants, or on our travels.

Laudan, Rachel (September–October 2010). "In Praise of Fast Food". UTNE Reader. Retrieved 2010-09-24. For our ancestors, natural was something quite nasty. Natural often tasted bad. Fresh meat was rank and tough, fresh fruits inedibly sour, fresh vegetables bitter.

Laudan, Rachel (September–October 2010). "In Praise of Fast Food". UTNE Reader. Retrieved 2010-09-24.

Levenstein, H, "Paradox of Plenty", pages 106–107. University of California Press, 2003.

Levinstein, Harvey. Paradox of Plenty: A Social History of Eating in Modern America. Berkeley: University of California, P. 2003. 228–229.

Lutz, Carroll; Przytulsk, Karen (2010). Nutrition and Diet Therapy. p. 257. ISBN 0803625235

Luxenberg, Stan. Roadside Empires: How the Chains Franchised America. New York: Viking, 1985.

Manohar. SS, BV Balasubramanyam and C Sheshachala, 2005, Convenience foods-Growth and prospects, Indian Food Industry, Vol 24, Page:56–57

Marketing Nutrition: Soy Functional Foods, Biotechnology, and Obesity, (2007), Brian Wansink, Champaign, IL: University of Illinois Press Helm & Gritsch, 2014.

McGinley, Lou Ellen with Stephanie Spurr, Honk for Service: A Man, A Tray and the Glory Days of the Drive-In. St. Louis: Tray Days Publishing, 2004. For photos of the Parkmoor Restaurants see Drive-In Restaurant Photos

Michael Pollan, 'Some of my Best Friends are Germs', New York Times Magazine, 15 May 2013, http://www.nytimes.com/2013/05/19/magazine/say-hello-to-the-100-trillion-bacteria-that-make-up-your-microbiome.html?hp

National Academies Institute of Medicine, Sodium Intake in Populations. http://www.iom.edu/Reports/2013/Sodium-Intake-in-Populations-Assessment-of-Evidence.aspx

Nikolova, HD, and Inman, JJ (2015). Healthy Choice: The Effect of Simplified Point-of-Sale Nutritional Information on Consumer Food Choice Behavior. *Journal of Marketing Research (JMR)*, 52, 817–835. doi:10.1509/jmr.13.0270

Payne, Stephanie. "Food processing". www.ikausa.com. Retrieved 14 November 2014.

Pietrangelo, Ann; Carey, Elea. "13 Effects of Fast Food on the Body". Healthline. Retrieved March 20, 2016.

Pollan, M (2009). In Defense of Food: an Eater's Manifesto. New York City: Penguin

Pomeranz, JL, & Adler, S (2015). Defining Commercial Speech in the Context of Food Marketing. *Journal of Law, Medicine & Ethics*, 4340–43 4p. doi:10.1111/jlme.12213

Potter, N.N. and J.H. Hotchkiss. (1995). "Food Science", Fifth Edition. New York: Chapman & Hall. pp. 478–513.

Psychol Bull, 129 (2003), pp. 873–886.

Public Health Nutr, 17 (2014), pp. 2721–2729.

Publications, Harvard Health. "Red meat and colon cancer—Harvard Health". Retrieved 2016-08-16.

Remar, D, Campbell, J, and DiPietro, RB (2016). The impact of local food marketing on purchase decision and willingness to pay in a foodservice setting. Journal Of Foodservice Business Research, 19, 89-108. doi:10.1080/15378020.2016.1129224

Riva, Marco; Piergiovanni, Schiraldi, Luciano; Schiraldi, Alberto (January 2001). "Performances of time-temperature indicators in the study of temperature exposure of packaged fresh foods". Packaging Technology and Science. 14 (1): 139. doi:10.1002/pts.521.

Robertson, G. L. (2013). "Food Packaging: Principles & Practice". CRC Press. ISBN 978-1-4398-6241-4

Salokya, on 19 October 2012 (19 October 2012). "वाईवाई र नेबिको बसिकुट गुणस्तरवहीन—Mysansar". Mysansar.com. Retrieved 7 November 2012.

Samyang Foods [1]. Retrieved 4 July 2008.

Schlosser, Eric, *Fast Food Nation: The Dark Side of the All-American Meal*, Houghton Mifflin Company, 2001.

Schultz, Howard with Dori Jones Yang, *Pour Your Heart Into It: How Starbucks Built a Company One Cup at a Time*, Hyperion, 1999.

Selke, S, (1994). "Packaging and the Environment". ISBN 1-56676-104-2.

Selke, S, (2004) "Plastics Packaging", ISBN 1-56990-372-7.

Shaw, Randy. "Food Packaging: 9 Types and Differences Explained". *Assemblies Unlimited. Retrieved 19 June 2015.*

Sheth, Jagdish (2011). *Models of Buyer Behavior.* Marketing Classics Press. Chapter 13. p. 255. ISBN 1613110561.

Shubin John A and H Madeheim, Plant Layout, New Delhi: Prentice Hall of India, 1986.

Social modeling of eating: a review of when and why social influence affects food intake and choice.

Solman, Gregory (March 19, 2008). "Nestle Eyes Dailey for Instant Breakfast". *Adweek. Retrieved 9 November 2013.*

Soroka, W (2009). "Fundamentals of Packaging Technology". Institute of Packaging Professionals. ISBN 1-930268-28-9

STAC - Project Information Center - 03-STAC-01 - Western U. S. Food Processing Efficiency Initiative.

Stillwell, E. J, (1991) "Packaging for the Environment", A. D. Little, 1991, ISBN 0-8144-5074-1.

Talwar, Jennifer (2003). *Fast Food, Fast Track: Immigrants, Big Business, and the American Dream. Westview Press. ISBN 0-8133-4155-8.*

Taneja, Staish and Gupta SL, Entrepreneurship Development and New Venture Creation: Galgotia Publishing Co 2001.

USAID. *Fortification Basis. Instant Noodles: A Potential Vehicle for Micronutrient Fortification.* Retrieved from http://www.dsm.com/en_US/nip/public/home/downloads/noodles.pdf

USDA

USDA (6 Nov 2010). "Commercial Item Description Soup, Noodle, Ramen, Instant" (PDF). USDA. Retrieved 16 Dec 2016.

Verma, JC, and Gurpal Singh, Small Business and Industry—A handbook for Entrepreneurs, New Delhi, Sage, 2002.

Ward, Elizabeth (2012). "Instant Breakfast". *Men's Health magazine.*

Warner, Melanie "Salads or No, Cheap Burgers Revive McDonald's." The New York Times April 19, 2006.

Wikipedia

Wongleedee, K. (2015). Marketing Mix and Purchasing Behavior for Community Products at Traditional Markets. *Procedia-Social and Behavioral Sciences, 1977th World Conference on Educational Sciences,* 2080-2085. doi:10.1016/j.sbspro.2015.07.323

World food consumption patterns—trends and drivers.

Yam, K. L., "Encyclopedia of Packaging Technology", John Wiley & Sons, 2009, ISBN 9780-470-08704-6.

Zhu Wenqian (March 23, 2015). "Chinese consumers lose taste for instant noodles".

Index

Acidity regulators 158
Acidulants 156, 159
AGMARK 372, 373
Air precooling 382
Alsi 116
Ambient environment 200
Angel investors 316, 319
Animal derived products 382
Animal-powered 241, 243
Anthocyanins 114, 164
Anticaking agents 159
Anti-foaming agent 161
Antioxidants 113, 151, 158, 161, 179
Anti-theft 198
Artificial colouring 162, 163
Artificial colours 81
Avante garde 356
Avenanthramides 137

Barrier protection 197
Beta testing 3
Biopsychosocial force 80
Bleached paper 206
Blind testing 15
Branded testing 15
Bulking agents 112, 161
Business finance 316

Canning 86, 87, 100, 165, 216
Capital intensive 100, 316
Carotenoids 114, 118, 164
Catechins 114
Cellulose 158, 203, 204, 223, 224
Chia 117
Chromolithography 234
Cloth materials 203, 222
Cod 115
Colour retention agents 164

Colourings 161, 163, 164
Combined layout 311
Commodity specific 110
Competitive pricing 342, 343
Composite containers 203
Consumer research 10–14, 18–20, 210
Consumer testing 5, 262, 263
Consumption per capita 64, 66, 68
Containment 195, 197, 392
Convenience foods 50, 76, 93, 100,
    102, 103, 105–107, 109, 119, 120,
    390, 393
Cooling 91, 177, 202, 256, 258, 375,
    376, 381–384, 386
Cost-plus pricing 342
Cup noodles 142, 145, 146

Debt financing 320
Demand pricing 342
Demographic 24, 29, 42, 53, 65, 78,
    79, 306, 352, 364
Dependent variables 7
Descriptive tests 266
DHA 115
Diaries 21, 25
Diet pattern shifts 67
Difference tests 266
Digital press 235
DPA 115

E number 152
Edible films 203, 220
Ellagic acid 114
Emulsifiers 152, 155, 157, 158, 165,
    170, 174
Enrichment 94, 155
Entrepreneur 31, 126, 303, 306, 307,
    310

Entrepreneurship 7, 302–304, 392, 395
Equipment 95, 96, 99, 101, 102, 180, 183, 187, 188, 214, 217, 221, 224, 233, 251, 254, 259, 282, 285, 286, 307–309, 313, 315, 316, 321–323, 326–333, 336–338, 343, 346
Equity 316, 317, 318, 319, 322, 324
Equity financing 317, 325
Ethos 354
Ethylene 205, 213, 214, 222, 225, 226, 378, 384
European union 64, 83, 117, 157, 168
Evaluation phase 188, 191

Fast food 122, 123
FCS 68, 69, 70, 71, 72
Fecula 175
Fish liver oils 115
Fixed expenses 343
Fixed position layout 310
Flavonoids 114, 118, 388
Flavourant 166
Flavourings 6, 115, 137, 145, 167
Flaxseed oil 117
Flaxseeds 116
Flexography 235
Focus groups 21, 25
Food additives 95, 119, 150, 152, 153, 155–157, 159, 164, 166, 169, 172, 173, 177, 178, 332, 367, 368, 393
Food advertising 352
Food carriers 251
Food choice questionnaire 42
Food colours 152
Food consumption trends 52
Food courts 126
Food deterioration 298, 300
Food group weights 71
Food groups 54
Food marketing 360
Food processing 73, 86, 87, 88, 89, 93, 95, 97, 99, 100, 102, 110, 112, 168, 305, 306, 338, 339, 360, 366
Food quality 7, 8, 9, 51, 91, 222, 262, 275, 277, 284, 285, 287, 288, 292, 299
Food security 59, 60, 72, 375, 376, 381
Food standards 372

Food transportation 249, 250
Fortification 94, 155, 381
Freight 243–248, 256–257, 344
Fssai 145, 367
Functional foods 109–111, 390

Gellants 175
Generic 110
Glass 203, 208, 209, 210, 211
Glassine paper 207
Glazing agents 171, 172
Global food consumption trends 64
Gluten 110, 143, 144, 170, 171
Gluten-free 110
GRAS 152
Gravure 235, 237
Greaseproof paper 206
Group I 152
Group II 152

Health claims 110, 359
Hemicelluloses 204
Human-powered 241, 242
Humectants 156, 172, 173, 174
Hydrocolloids 144
Hydrocooling 383
Hygiene 35, 97, 98

Idea generation 11
Independent variables 7
Index 38, 261, 287, 288, 293, 295, 345
Instant mashed potatoes 148
Instant noodles 142, 145, 393, 395
Instant oats 134
Instant rice 141
Instant soup 146
Insulated containers 251
Insurance policies 317
Interviews 3, 14, 16, 21–23, 27
IPOs 320
Isoflavones 118

Kraft paper 206

Lanolin 172
Lignin 111, 204
Linoleic acid 117
Linolenic acid 117
Lithography 233, 234

Loans 304, 316, 317, 318, 321, 323, 324, 325
Location layout 309, 311
Logos 354
Long-haul transport 247
Louis pasteur 87

Maggi 145
Marine oils 116
Market research 11, 18–20, 26, 28, 29, 349, 351, 360
Marketing analysis 11
Marketing management 10
Marketing research techniques 28
Markup pricing 343
Media mix 351, 352
Metal 79, 95, 96, 158, 195, 203, 211, 216, 217, 220, 221, 222, 223, 224, 226, 227, 228, 230, 237, 250, 293, 296, 333, 383
Microbial gut flora 95
Migration 202
Minimarket-testing 16
Mixture problems 8
Momofuku ando 142, 147, 390

Natural additives 151
New product development 11
Nicolas Appert 87
Nutrition labelling 111, 132
Nutrition transition model 78, 109

Objective evaluation 284
Offset printing 236, 237
Overhead expenses 343

Pasteurization 87
Pathos 353
Persuasive techniques 353
PFA 366
Popkin 78
Population growth 65, 66, 68
Preservatives 87, 119, 144, 152, 153, 155, 156, 157, 161, 178, 179, 219
Projective techniques 21, 25
Protected designation of origin (PDO) 83

Qualitative research 16
Quantitative research 11, 27
Questionnaires 21, 22, 23, 24
Quick cooking 134

Rail 241, 243
Ramen 142, 145, 395
Rating tests 266, 268
Ready to cook (RTC) foods 108
Ready to drink (RTD) beverages 109
Ready to eat (RTE) foods 107
Ready to fry (RTF) foods 108
Ready to serve (RTS) beverages 109
Ready to use (RTU) foods 108
Recipe problem 8
Refrigeration 100, 116, 148, 178, 252, 254–259, 383, 384, 389
Regenerated cellulose 203, 223
Regression analysis 15
Resveratrol 114
Retort processed foods 108
RFID tags 198, 238
Rice bran 112, 113
Rolled oats 135, 136, 137

Screen printing 234
Secondary package 195
Sensory panel 262
Shelf life 89, 90, 92, 102, 103, 109, 142, 172, 173, 174, 197, 211, 218, 219, 225, 230, 231, 233, 256, 376, 382
Simulated test markets 16
Simulations 15
Standardized recipes 180, 181, 182
Steel-cut oats 135
Synthetic additives 151

Temperature control 250, 251, 257, 259, 378
The marketing mix 30

Variable expenses 344
Venture capital 304, 318

Warrant 118, 174, 320
Weasel words 355
Wooden containers 203